Cambridge Studies in the History of Psychology

GENERAL EDITORS: WILLIAM R. WOODWARD AND MITCHELL G. ASH

Constructing the subject

Constructing the subject

Historical origins of psychological research

Kurt Danziger
York University

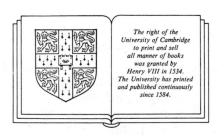

*The right of the
University of Cambridge
to print and sell
all manner of books
was granted by
Henry VIII in 1534.
The University has printed
and published continuously
since 1584.*

Cambridge University Press

Cambridge
New York   Port Chester   Melbourne   Sydney

Published by the Press Syndicate of the University of Cambridge
The Pitt Building, Trumpington Street, Cambridge CB2 1RP
40 West 20th Street, New York, NY 10011, USA
10 Stamford Road, Oakleigh, Melbourne 3166, Australia

First published 1990

Printed in Canada

Library of Congress Cataloging-in-Publication Data
Danziger, Kurt.
Constructing the subject : historical origins of psychological
research / Kurt Danziger.
p.   cm. – (Cambridge studies in the history of psychology)
ISBN 0-521-36358-6
1. Psychology, Experimental – History.   I. Title.   II. Series.
BF181.D27   1990
150'.72 – dc20   89–22160
                                                          CIP

British Library Cataloguing in Publication Data
Danziger, Kurt
Constructing the subject : historical origins of
psychological research. – (Cambridge studies in the
history of psychology).
1. Psychology, history
I. Title
150'.9

ISBN 0-521-36358-6 hard covers

BF
181
.D27
1990

Contents

v

Preface

Historical studies of the sciences tend to adopt one of two rather divergent points of view. One of these typically looks at historical developments in a discipline from the inside. It is apt to take for granted many of the presuppositions that are currently popular among members of the discipline and hence tends to view the past in terms of gradual progress toward a better present. The second point of view does not adopt its framework of issues and presuppositions from the field that is the object of study but tends nowadays to rely heavily on questions and concepts derived from studies in the history, philosophy, and sociology of science. A history written from the insider's point of view always conveys a strong sense of being "our" history. That is not the case with the second type of history, whose tone is apt to be less celebratory and more critical.

In the case of the older sciences, histories of the second type have for many years been the province of specialists in the history, philosophy, or sociology of science. This is not, or perhaps not yet, the case for psychology, whose history has to a large extent been left to psychologists to pursue. Accordingly, insiders' histories have continued to have a prominence they have long lost in the older sciences. Nevertheless, much recent work in the history of psychology has broken with this tradition. Although my own formal training was in psychology, I have tried, in the present volume, to adopt a broader perspective and to analyze the history of this discipline within a framework that owes much to the field of science studies. Inevitably, the result is not what it would have been had my own disciplinary affiliation been outside psychology. Perhaps I have produced a different kind of insider's history. That is for the reader to judge.

This book is not, however, a history of modern psychology. First of all, I have limited myself to certain aspects of that history, albeit aspects that I consider to be fundamental. In the first chapter I explain my reasons for my choice of topics. The general framework I have adopted is of course applicable to other topics beyond those analyzed here.

The time period covered by the present study is also limited. Because the focus is on historical *origins* of psychological research, developments beyond the first four decades of the twentieth century are not considered. This work is concerned with the formative period of modern psychology, and the beginning of World War II is taken as a rough marker of the end of that period. This does not exclude brief references to later developments, nor does it include some developments at the end of this period whose major impact did not occur until somewhat later.

Finally, practical considerations made it impossible to extend the analysis presented here to many important fields whose early development was peripheral to the major thrust of the discipline. For example, when I first planned this book I had intended to include some discussion of psychoanalysis. However, I soon realized that it would require a different major study to do justice to that topic. Somewhat similar considerations apply to other potentially interesting areas like developmental and social psychology.

I am indebted to my graduate students, Adrian Brock, John Dunbar, Cindi Goodfield, James Parker, Nancy Schmidt, Peter Shermer, and Richard Walsh, some of whom did much of the content analysis of psychological journals that I have used in various parts of this book and all of whom helped me to clarify my ideas in countless discussions. James Parker and Peter Shermer made specific contributions of their own that strongly influenced my treatment of some of the topics in chapters 5, 9, and 10.

I would also like to express my thanks to Mitchell Ash, Gerald Cupchik, Gerd Gigerenzer, Rolf Kroger, Ian Lubek, Jill Morawski, David Rennie, Irmingard Staeuble, Piet van Strien, Malcolm Westcott, and Linda Wood for many valuable suggestions and comments. By advice and by example they have improved this book, though they are in no way responsible for its contents and deficiencies.

Financial support from the Canadian Social Sciences and Humanities Research Council made possible the research on which this book is based. I appreciate this assistance as well as that received from York University, Toronto, which supported this project generously in the form of research funds and a Faculty of Arts Fellowship, which gave me sufficient time to complete the manuscript.

In some parts of this book I have drawn on previously published material. In particular, I would like to acknowledge the permission granted by Yale University Press to use a modified version of my contribution to a volume edited by Jill Morawski, entitled *The Rise of Experimentation in American Psychology*. This appears as chapter 6 of the present volume. Table 5.2 is also taken from this source. Acknowledgment is also due to the American

Psychological Association and the Clinical Psychology Publishing Company for permission to use shorter excerpts from papers previously published in the *American Psychologist* and the *Journal of the History of the Behavioral Sciences*. The last section of chapter 8 is a slightly revised version of part of a chapter that appears in M. G. Ash and W. R. Woodward, eds., *Psychology in Twentieth Century Thought and Society* (Cambridge: Cambridge University Press, 1987).

Last, but by no means least, I want to express my appreciation to Judy Manners for the extraordinary skill and patience she displayed in typing the book manuscript.

Kurt Danziger
Toronto

1

Introduction

What this book is about

What exactly constitutes a field like scientific psychology? Is it constituted by its most innovative and influential contributors; by the scientific findings that it has produced; by the theories it has elaborated; by its concepts, techniques, or professional associations? Obviously, all this and more goes into the making of a field, but most of us would probably see some of these components as playing a more essential role than others. Even if we refuse to commit ourselves explicitly we are likely to imply that certain components define the field more effectively than others by the way we organize our knowledge. For example, in the systematic presentation of information derived from the field of psychology or one of its parts, the material is most commonly organized in terms of prominent contributors, important findings, or influential theories. A perhaps unintended message of such communications is that psychology *is* its theories, *is* its findings, or *is* its individual contributors.

The way in which we organize a field will determine the way we organize its history. If we see the field of psychology as essentially an aggregate of individual contributors, we are likely to treat the history of the subject in terms of a succession of prominent figures. If psychology is its theories or its findings, then its history will become a history of psychological theories or psychological findings. Our organization of the history of the field will also serve as a subtle justification of the way we have characterized the field in the present.

Most psychologists have been taught to characterize their own scientific activity in terms of a framework that is derived from nineteenth-century physical science. They see themselves as individual investigators who seek to accumulate facts about some aspect of nature by the use of appropriate

1

hypotheses and techniques. When they describe the historical development of their field they are apt to do so in much the same terms, representing it as a succession of individual contributors who accumulated "findings" on the basis of progressively refined hypotheses and increasingly sophisticated instrumentation. Of course, the sad truth sometimes forces a departure from this framework, but the framework continues to operate all the same.

To put it very simply, this book is about some crucial elements that are missing in this framework. What is missing is the recognition of the socially constructed nature of psychological knowledge. The received view is based on a model of science that is reminiscent of the tale of Sleeping Beauty: The objects with which psychological science deals are all present in nature fully formed, and all that the prince-investigator has to do is to find them and awaken them with the magic kiss of his research. But in truth scientific psychology does not deal in natural objects. It deals in test scores, rating scales, response distributions, serial lists, and innumerable other items that the investigator does not just find but constructs with great care. Whatever guesses are made about the natural world are totally constrained by this world of artifacts.[1] The same holds true for the immediate human sources of the psychologist's information. The psychologist's interaction with such sources takes place within a well-regulated social role system, and such roles as that of experimental subject or of client in therapy are the direct result of the psychologist's intervention.

In talking about a field like scientific psychology we are talking about a domain of constructions. The sentences in its textbooks, the tables and figures in its research reports, the patterned activity in its laboratories, these are first of all products of human construction, whatever else they may be as well. Although this seems quite obvious, certain implications are usually evaded. If the world of scientific psychology is a constructed world, then the key to understanding its historical development would seem to lie in those constructive activities that produced it. But this insight has not guided many historical studies. In the past the effects of a naive empiricism may have assigned an essentially passive role to investigators, as though they merely had to observe or register what went on outside them. But this is no longer a popular position.

In more recent times the well-known contrast between "context of discovery" and "context of justification" gave expression to a pervasive tendency to relegate the necessary subjective component in scientific activity to a mysterious underworld that was not susceptible to logical analysis.[2] So there grew up a strange duality in the historiography of fields like psychology, where one kind of historical review would restrict itself to the logical succession of hypotheses and evidence while the second kind would describe the personal lives of those individuals who were the authors of the hypotheses and the producers of the evidence. Whereas the first kind of review implied that scientific progress was an affair governed by purely rational considerations, the second kind picked up the irrational component and located it in the personal quirks and accidental events that characterized the lives of historically important contributors. The two approaches were the product of a tacit consensus about the fundamental split between two components of scientific activity: a rational transindividual component, of primary importance in the context of justification; and an irrational individual component, important in the context of discovery.

What is missing from this account is any appreciation of the fundamentally social nature of scientific activity.[3] What unites individual contributors is not simply their common possession of the same logical faculties and their common confrontation of the same external nature. Their social bonds are a lot more complex than that. They are related by ties of loyalty, power, and conflict. They share interests as well as logical faculties, and they occupy positions in wider social structures. In this social world of science the neat distinction between rational and irrational components crumbles. The fundamental issue in research is not whether the lone investigator can verify his hypotheses in the privacy of his laboratory but whether he can establish his contribution as part of the canon of scientific knowledge in his field.[4] In other words, the issue is one of consensus, and consensus is not entirely a matter of logic. It involves prior agreements about what is to count as admissible evidence and shared commitments to certain goals. It involves vested interests and unexamined biases.

Once we recognize the essentially social nature of scientific activity, we are compelled to see both the "context of discovery" and the "context of justification" in a different light. The context of discovery is in fact a context of construction, of theories, of instruments, and also of evidence. For the data that appear in the pages of psychological journals are no less a product of the constructive ingenuity of their authors than are the instruments and the theoretical hypotheses; they are not raw facts of nature but elaborately constructed artifacts.[5] However, these artifacts are constructed according to explicit rational schemes accepted within a certain community of investigators.[6] That is why there is a real history of psychological research practice that is neither a series of narratives about famous psychologists nor an enumeration of their successive "findings." This history involves the changes that have taken place in the constructive schemes that psychologists have used in the production of those objects that form the accepted content of their discipline.

But what is meant by these "constructive schemes"? In the first place, such schemes are not just cognitive frameworks for the *interpretations* of empirical data but involve practical rules for the *production* of such data. It is true that general concepts and theories also function as constructive schemes that give a particular meaning to the objects in which the discipline deals. But such interpretive schemes are found in purely speculative psychologies as well as in empirical psychology. Psychology underwent a fundamental change toward the end of the nineteenth century when its practitioners became decisively committed to specific practical methods of data production. The application of these methods became the special characteristic of the field and distinguished it from everyday psychology as well as from its own intellectual predecessors.[7] By the use of these practical methods, modern psychology created a new world of psychological objects that increasingly defined the field and to which any purely theoretical developments were forced to accommodate.

This emphasis on practical constructive schemes does not however lead simply to a history of psychological methods in the conventional sense. The difference lies in the way one conceives of method. Conventionally, empirical methods in psychology are conceived simply as tools for the achievement of certain technical goals. Thus, they need be evaluated only in terms of a logic of means and ends, where the ends are taken as existing independently of and prior to the means. For instance, we may choose the measurement of a particular psychological quality, such as intelligence, as an end, and then trace the effectiveness of various instruments for achieving this end. Or we may take it as an end that we want to assess the simultaneous contribution of a number of factors to psychological effects and then attribute the growing popularity of certain statistical techniques to their superior effectiveness in achieving this preexisting goal. It is possible to trace the history of psychological methodology in terms of such a purely instrumentalist framework, but that is not what is being attempted here.

The concept of investigative practice is wider than the concept of methodology. As conventionally understood, the latter involves an abstraction of certain rational and technical aspects of investigative practice from all the other aspects. The practice of investigators is treated as though it consisted only of logical and technical operations performed by independent individual investigators on bits of the natural world. Left out is the fact that investigative practice is very much a social practice, in the sense that the individual investigator acts within a framework determined by the potential consumers of the products of his or her research and by the traditions of acceptable practice prevailing in the field.[8] Moreover, the goals and knowledge interests that guide this practice depend on the social context within which investigators work. Finally, in psychological research there is the additional important consideration that the investigator is not the only human participant whose actions are necessary for the practice of investigation to proceed. Unless the psychologist works with animals, he or she will also require the collaboration of human sources of data without whose contribution there would of course be nothing to report.

The notion of investigative practice then involves the social dimensions of research activity as much as the logical ones. The latter are recognized as being embedded in a social matrix that includes such factors as the pattern of social relations among investigators and their subjects, the norms of appropriate practice in the relevant research community, the kinds of knowledge interests that prevail at different times and places, and the relations of the research community with the broader social context that sustains it.

As long as we limit our conception of psychological research practice to its purely rational aspects, we will be inclined to think of the history of that practice solely in terms of technical progress. The norms of good scientific practice will be seen as belonging to an unchangeable transhistorical realm where eternal rational principles rule. All that is left for worldly history is the narrative of how clever investigators came to apply these eternal principles with ever greater concrete effectiveness to larger and larger bodies of knowledge. However, if we refuse to perform this rationalist reduction, we will find that in the history of psychological research practice the most significant changes were changes in the ends rather than improvements in the means.

Although methodological rationalism has adherents outside psychology, the doctrine has long played a particularly important role for this discipline. In the more established fields of natural science, where elaborate deductive procedures are part and parcel of theoretical discourse, theory as much as method was regarded as the repository of scientific reason. But in psychology theoretical constructions have seldom been marvels of logical sophistication, and pessimism about the likelihood of reaching rational consensus on the basis of theory has long been widespread. It is generally accepted that there will be controversy about theoretical fundamentals and that personal, cultural, and historical factors play important roles in the elaboration and acceptance of psychological theory. But this state of affairs is hardly compatible with the claims by the discipline for the objectivity of its insights into human behavior. Therefore, such claims in psychology have depended almost completely on the rational virtues of its methodology. It was only because of the logical–technical features of its investigative practice that psychology could give some plausibility to its claims for scientific status.

Investigative practice therefore constitutes an area of considerable anxiety within the discipline of psychology.[9] Concern with questions of methodological orthodoxy often takes the place of concern about theoretical orthodoxy when research or its results are discussed and evaluated. These preoccupations with the purity of method frequently deteriorate to a kind of method fetishism or "methodolatry."[10] From this point of view there may be something distinctly subversive about the suggestion that the sphere of methodology is not a realm of pure reason but an area of human social activity governed by mundane circumstances like any other social activity.

Nevertheless, the consequences of this suggestion should be explored, for not to do so exposes one to all the risks entailed by a naive and self-deluded style of scientific practice.

Once we restore the abstraction of a purely rational methodology to the broader context of investigative practice, it becomes possible to see it as the primary medium through which social forces have shaped the production of the objects of scientific psychology. But this general pattern can only be fully appreciated when studied in its historical development. Before we can undertake this task, however, we require some clarification of the sociological and historiographic aspects of the topic.

The social generation of scientific knowledge

In recent years references to science as a social activity have become commonplace, and there is a rapidly growing field devoted to the social study of science. Whereas the earlier sociology of science was more a sociology of scientists and exempted the content of science from its purview, more recent developments in this field have pointed in the direction of a genuine sociology of scientific knowledge.[11] In other words, scientific knowledge has increasingly been seen as a product of certain quite specific social processes, and many of its features have been studied in terms of their dependence on these processes. With a few notable exceptions,[12] psychological knowledge has been largely exempt from this process. This may be due to the existence of two large bodies of opinion: For one the social dependence of psychological knowledge is too obvious to be worth studying, whereas for the other the whole topic is too threatening to psychology's hard-won scientific status to be taken seriously. The fact is, however, that psychology's social practices of investigation are too tied up with its claims to scientific status to be ignored in that context. Still, our understanding of the social determination of psychological knowledge must remain rather abstract until we understand the crucial mediating role played by investigative practice.

It may be helpful to think of the social contexts of investigative practice in terms of three concentric circles of varying radius. (Figure 1.1) The innermost circle represents the immediate social situation in which the information that will later be transformed into research data and scientific psychological knowledge is generated. The next circle represents the research community by which the product of the investigation must be accepted if it is to count as scientific psychological knowledge. The outer circle stands for the wider social context within which the research community is embedded – the sources of research support, professional institutions, potential external consumers of knowledge products and skills, representatives of prestigious disciplines, and so forth.

The image of concentric circles is of course meant to represent the idea

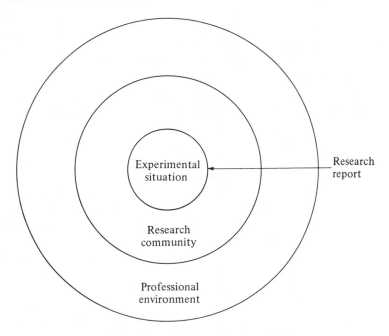

Figure 1.1. Social contexts of investigative practice

that both the immediate investigative situation and the research community are to be regarded as embedded in social relationships that extend beyond them. Similarly, a truly comprehensive account would have to add further circles beyond the three discussed here to represent even broader cultural, political, and economic contexts. But because our concern is specifically with investigative practice, we will limit ourselves to the inner three circles, considering in turn some of the more important questions that arise within each.

At the innermost level we have to deal with what are essentially face-to-face situations. Before the whole process of producing material that counts as scientific psychological knowledge can begin, some individuals have to get together to generate the information that forms the starting point for this process. In coming together for this purpose they set up certain social situations with a distinct character – the situations of mental testing, psychological experimentation, or interviewing, to mention the most obvious ones. In the case of the first two, which are by far the most important ones in psychology, the social interaction of the participants is partly mediated by various kinds of hardware and software, mental-test materials, and laboratory apparatus in particular. This has permitted investigators to overlook the social character of all these investigative situations and to pretend that their subjects reacted only to the dead materials presented to them and not to those who did the presenting. Nevertheless, it is an elementary truth that the gathering of psychologically relevant information requires the active social participation of individuals who will act as the source of this information. Whatever else it may be, the psychological experiment or test is therefore a social situation and as such it must share the characteristics that are found in all social situations.

Of course, there are marginal cases where the social status of the investigative situation is less straightforward, such as the use of documentary sources or the case where individuals experiment on themselves. But neither of these has accounted for more than a tiny fraction of psychological research over the past century. Their rarity only serves to emphasize the inherently social nature of most psychological investigations.

Until relatively recently the total blindness of psychological investigators to the social features of their investigative situations constituted one of the most characteristic features of their research practice.[13] However, the recognition of so-called experimenter expectancy effects and demand characteristics has now become commonplace, and there is a large empirically oriented literature on "the social psychology of the psychological experiment."[14] Although this undoubtedly represents an advance on psychologists' traditional naiveté in regard to such issues, the limitations of this literature are more remarkable than its achievements from the viewpoint of furthering an understanding of the social process of generating psychological knowledge.

In the first place, much of the literature on social effects in psychological experiments is devoted simply to demonstrating the mere existence of such effects without being directed at furthering an understanding of the social processes involved in psychological investigation.[15] Such conceptualization of social process as does occur in this work is usually confined to social *psychological* rather than sociological categories and therefore does not address the fact that the responses of individuals to experimental situations take place in the context of a miniature social system in which both experimenters and their subjects participate. Moreover, there is a pervasive tendency to relegate the social aspects of psychological experimentation to the status of "artifacts" or, in other words, disturbances of the process of research that do not belong to its essential nature.[16] Thus the rational abstraction of a purely logical, asocial, and ahistorical research process remains inviolate and apparently beyond the reach of empirical correction. Accordingly, in this research program the preferred method for studying the social aspects of experimentation is the experimental method itself, an enterprise whose results are difficult to interpret without getting mired in an infinite regress.[17]

Such problems mostly derive from a single source, namely, the pretense that psychological experiments are not in principle different from experiments in the natural sciences.[18] Because in the latter case experimenters are able to treat whatever they are investigating purely as a natural object, it is believed that human data sources must be treated in the same way if psychology is to function as a proper experimental science.[19] But human subjects in psychological experiments are in fact unable to behave simply as natural objects. Even if they try to do so, which depends entirely on their appraisal of the social situation they are in, they negate this fictional goal in the very act of trying to reach it, because such efforts represent an exercise of their social agency.[20] Psychological experiments are therefore different in principle from experiments in physics because the experimenter and the human data source must necessarily be engaged in a social relationship. This is no "artifact" but one of the essential preconditions of having a viable experimental situation. Consequently, experimental results in psychology will always be codetermined by the social relations between experimenters and subjects. The precise significance of this factor may be variable, but that it is always present is not a matter open to doubt.

Of course, the old notion that the knowledge produced by experiments depends only on the interaction of individual experimenters with the materials of nature is false even in the case of the true natural sciences. We now understand a great deal about how the kind of knowledge gain that results from an experiment depends quite crucially on the interaction of scientists with each other.[21] But this is a level of social influence that psychological research shares with research in other experimental sciences. Beyond this there is however a level of essential social interaction peculiar to psychology experiments – the interaction of experimenters with the human sources of their data. It is to be hoped that after a century of lusty life, experimental psychology now feels sufficiently secure to face the fact that there may be essential differences between itself and the older experimental sciences. The crude doctrine that relegated everything not cast in the image of experimental physics to the category of the "mystical" is being increasingly treated as a historical curiosity.

Finding out more about the social relationships that exist between experimenters and their human subjects is surely better than pretending they are not there or are of no importance. Now, there are different ways of pursuing such a goal. The experimental way, which has already been alluded to, tends to be limited to investigating certain *social psychological* factors that operate in psychological experiments.[22] But psychological factors like the experimenter's expectancies or the subject's "evaluation apprehension" operate within a certain social framework that has to be taken for granted in such studies. This framework is provided by the traditions and conventions of psychological experimentation that have developed over the years and are now well understood by all experimenters and by most of their subjects. In those societies in which it is practiced on any scale, the psychological experiment has become a social institution recognized by most people with a certain level of education. As in all social institutions the interaction of the participants is constrained by institutional patterns that prescribe what is expected and permitted for each participant. The successful conduct of psychological experiments depends on the willingness of all participants to abide by these rules, and that of course can only happen if they have the appropriate level of previous knowledge to understand the rules. We know that in cross-cultural research this cannot always be taken for granted.

But before we discuss this as a third-world problem we might remember that hardly more than a century ago the institution of the psychological experiment was as unknown everywhere as it might now be in parts of the third world. It is legitimate to ask how this social construction ever arose in the first place and how it ended up taking the form we are so ready to take for granted today. Like all social institutions, the psychological experiment not only has a certain social structure that can be analyzed but a history that can be traced. Ultimately, this institution is part of the history of those societies that produced it and can be expected to bear the marks of its origins. Thus, if we wish to improve our understanding of the accepted social framework within which individual participants in psychological experiments have to function, we will have to adopt a historical rather than an experimental approach. For neither experimenters nor their subjects enter the investigative situation as social blanks to be programmed in an arbitrary manner. Both are the products of a distinct historical development that has left a heavy sediment of blind faith and unquestioned tradition. It is precisely this historical development that we will attempt to trace in the present volume.

Although studies of the social framework for the interaction of investigators and their subjects are virtually nonexistent, we move onto relatively more familiar ground when we consider the two outer circles that determine the social practice of research. In their interaction with each other and with a wider social context, psychological investigators face problems that are not different in principle from the social problems faced by other groups of scientists. From this perspective we recognize the social practice of scientific psychological investigation as involving a certain organization of work for the purpose of generating a certain kind of product that is identified as psychological knowledge. This identification of the product is accomplished by what amounts to a system of certification on the part of recognized authorities working within an established institutional framework.[23] The kind of knowledge the scientific worker is after is not private but public knowledge.[24] That means it cannot be achieved without acceptance by some kind of public. If the knowledge is to bear the stamp of science, the public by whom it is accepted must be a scientific public.

Psychological research is not something that is carried on by individual investigators working on their own account. Nor are its social aspects limited to the interaction between investigators and the human sources of their data. In designing their experiments and in preparing the cognitive yield of their investigations for publication, researchers must be constantly mindful of the acceptability of their product to a particular public. That acceptability depends on whether the product measures up to currently prevailing standards of what constitutes scientific psychological knowledge. The fact that these standards have been thoroughly internalized by most investigators in the course of their professional socialization does not in any way affect the public status of these standards. What this means is that the scientific community is intimately involved in the social practice of scientific investigation.

As in the case of experimenter–subject interactions, there are two different levels on which the analysis of this aspect of social practice can be pursued. There is the social psychological level where one would be asking questions about how individual investigators adapt to prevailing standards of scientific knowledge and how they negotiate different interpretations of these standards. But at this level the existence of the standards themselves would have to be taken for granted. The systemic or historical level of analysis would treat precisely this existence of current standards as problematic and would attempt to trace their historical origin and development. This level of analysis will be pursued here.

Some historiographic considerations

The fact that currently dominant standards of what is to count as scientific psychological knowledge have not always existed and were not always dominant is one that is universally recognized. Every well-trained research worker in the field knows that the very idea of a scientific psychology only arose in relatively recent times and that its practical enactment belongs to an even more recent period. It is also well known that subsequent to the founding of experimental psychology there was a major controversy about the standards of what was to constitute scientific psychological knowledge. This controversy is usually represented as one between introspectionists and behaviorists, though, as we will see later, the story is more complex than that. What is worthy of note, however, is that an arch introspectionist like E. B. Titchener always justified his investigative practice in the name of *science* and denigrated the practice of his opponents as being not science but technology.[25] The more naive textbook treatments describe this controversy simply as a conflict between truth and error, the implication being that there can be only one conceivable true version of what constitutes scientific psychological knowledge. But if we rid ourselves of the irrational presumption that truth will necessarily vanquish error, and that therefore whatever achieves historical dominance must be truer,[26] this controversy becomes one between rival conceptions of what was to count as scientific psychological knowledge.

In studying historical changes in prevailing conceptions of psychological knowledge, the question of the relative truth of these conceptions is not relevant. First of all, we are dealing with standards, criteria, and ideals rather than with matters of fact, and so the attribute of truth is simply an inappropriate one. The introspectionists' choice of a different kind of knowledge was a matter of preference, not a matter of error. Whether what they did was scientific depends on one's definition of science. If their definition turns out to be different from that of the behaviorists, this again is a question of preference, which can only be seen as a matter of being right or wrong if a particular definition of science is accepted as the only true one in some absolute, ahistorical sense.[27]

But such ahistorical standards usually turn out to be the particular standards that happen to be currently popular. Elevating them to the status of final criteria therefore involves the patent illusion that historical development has reached a kind of culmination in the present and will now stop. The conservative implications of this position need no elaboration. When we attempt to trace historical changes in the social practice of investigation, we have to put questions of truth on one side. For here we are not dealing merely with technical means whose logical and practical adequacy can be assessed in terms of some accepted goal but with the goals themselves. Particular scientific communities are characterized by their commitment to certain scientific goals and this commitment determines the nature of their members' investigative practice.

What is certainly relevant in this context are the reasons for existing commitments to certain ideals of scientific practice and the reasons for changes in these commitments. These reasons are to be found in the common historical situation faced by members of a particular scientific community. Such communities never exist in a social vacuum, of course, but find themselves under the necessity of adopting positions in relation to other groups of investigators, to those who control the material resources for research, and to the general lay public. There may be other groups of investigators who are particularly well established and influential and who serve as models of successful practice. Those who ultimately control social resources must be persuaded to divert some of them to particular groups of investigators rather than to other purposes. At all times a newly emerging discipline like scientific psychology had to be careful to distinguish its cognitive product from the everyday knowledge of lay publics and from the rival claims of other disciplines. The requirements of potential consumers of a discipline's cognitive products will also pull their weight. All these and other similar factors help to shape the kinds of knowledge goals that prevail in a given field at a given time and therefore determine the dominant patterns of investigative practice.

This level of social determination can be thought of as constituting the outermost of the three concentric rings that were previously used to represent the multiple layers of the social practice of investigation. The innermost ring represents the direct social interaction of the participants in the research situation, and the middle ring represents the interaction of members of the scientific community with each other. Although each circle makes its own specific contribution to the overall pattern of investigative practice, these contributions are not autonomous but derive their significance from their embeddedness in a wider social context.

Any attempt at studying the historical development of investigative practice faces certain practical problems. The most significant of these problems arises from the fact that the realm of practice involves a great deal of that "tacit knowledge" that has long been recognized as playing a crucial role in scientific investigation.[28] In fact, the tacit component tends to be much larger on the level of practice than on the level of theory. Aspiring investigators are typically socialized into the craft by following examples and may never be required to account to themselves or to others for choosing procedures that have the weight of tradition behind them. Moreover, the accounting that they are required to do is typically limited to purely technical considerations. It is precisely the social aspects of scientific practice that are systematically excluded from practitioners' discussions about methodology. In psychology, where some of these aspects can no longer be overlooked entirely, they are discussed as "artifacts" that may disturb the normal rational course of investigative practice.

In other words, investigators tend to be poor direct sources of information when it comes to certain aspects of their practice. This would not matter too much if we wanted to study contemporary practice because we might be able to observe investigators actually at work or question them at length about their actions.[29] But if we want to understand current practice in terms of its historical development, we are precluded from proceeding in this way. Documentary evidence, where it deals explicitly with problems of investigative practice, is typically limited to the purely technical aspects. In particular, textbook discussions of methodological issues tend to suffer from this limitation. They only become historically interesting when changes in the treatment of certain issues are compared over time, an approach that permits the emergence of some fundamental questions whose answers are taken for granted in any single textbook presentation. But only certain aspects of investigative practice ever receive extensive textbook treatment, and for the rest we must look for other sources of evidence.

Methodological prescriptions by prominent representatives of the field can sometimes be of value. However, such statements tend to belong to the ideology rather than to the practice of research. They tell us what their authors thought they were doing or what they wanted others to think they were doing, but they do not necessarily tell us what they actually did. The significance of such statements always needs to be assessed against independent evidence about actual practice. In any case, the statements of individuals, no matter how prominent, may have little relationship to general practice, which is our ultimate interest. Maintaining a clear distinction between actual practice and discourse about practice is necessary because the latter involves the ways in which practitioners justify their practice to themselves and to others.[30] A form of social accounting, it provides only a very partial and often a distorted reflection of the practice as it might appear to a less partial observer with a broader perspective.

In regard to discourse about practice, it is difficult to escape the impression that its significance at different times in the history of modern psychology has varied widely. Such discourse was obviously crucial in the early years of experimental psychology, when the field was trying to define itself and to differentiate itself from related fields. In dealing with this stage, an understanding of prevailing discourse about practice is as important as an analysis of actual practice. But after the end of World War I, at least in the United States, psychology seems to have entered a more settled period during which answers to most of the fundamental methodological issues were increasingly taken for granted and discussion became limited to technical issues. (Nearer our own day there was a revival of interest in the more fundamental issues, but this takes us beyond the historical scope of the present volume.) Accordingly, we will pay less attention to discourse about method than to an analysis of actual practice. For, as we will see, actual practice underwent considerable change even though this was not always adequately reflected in current discourse about practice.

Fortunately, the conventions of scientific practice provide us with a veritable mine of information about those aspects that are ignored or distorted in discussions about technique. The products of research practice consist of reports published in special media that have a certain standing in some relevant research community. Such products constitute a large part of the crucial public domain of science. As such they are preserved for posterity and are therefore available for inspection as historical documents long after they have served their original purpose.[31] These published research reports in recognized scientific journals provide us with the historical record that enables us to trace significant changes in investigative practice over the years.

Given their potentially enormous informative value and their relatively easy accessibility, it is surprising that research reports have so rarely been subjected to systematic historical study.[32] Part of the reason for this may lie in the tendency to treat such reports at face value even though many years may have elapsed since their original publication. But a research article can inform us about much more than the specific data its authors wanted to communicate to their contemporaries. By the time that experimental psychology emerged, the conventions governing scientific research publications had developed to the point where they prescribed that in principle enough information about procedures had to be supplied to enable other investigators to replicate the published study. Of course, publication practices often fell short of this ideal, and in psychology it was in any case unrealistic because the historically unique nature of each study was accentuated by the use of human data sources. Nevertheless, the conventions of scientific publication exerted sufficient influence to ensure the inclusion of a considerable amount of information that bears on the development of investigative practices. For example, psychological research reports always contain some direct information about the identity of the participants in the research study and enough indirect information to enable one to draw some conclusions about the kind of relationship that must have existed between investigators and their human subjects.

The systematic analysis of research articles published in the major journals of the field will therefore form an important source of information for our historical study of the development of research practice. Much of this information is obtained by analyzing aspects of the research reports to which their authors would have attached little significance at the time. What these authors were interested in was the communication of their findings in an acceptable form; what we are interested in are, first, the tacit, taken-for-granted aspects of their social practice that enabled them to produce something they recognized as "data" in the first place, and second, the means they used to make these data presentable.[33] Thus our analysis will attempt to probe behind the conventional facade of the research article to the kinds of practices of which it was the product.[34]

To return once again to the representation of the social practice of investigation by three concentric rings, as developed in the previous section, the published research report can be thought of as existing on the boundary between the innermost and the middle ring. In one direction the research article reflects something of the social interaction among the participants in the research situation, but it is also directed at a particular scientific audience by which it hopes to be accepted. To be so accepted, it must present its information in such a way as to conform to certain standards of what constitutes acceptable knowledge in the field. An appropriate analysis of research articles can therefore provide us with some insight into the two levels of social construction that have gone into its production: first, the social structuring of the research situation in such a way as to yield a certain type of information, and second, the restructuring of this information to make it fit a certain prescribed model of what scientific knowledge in the field should look like.

But even if this kind of analysis were successful, it would yield at best a static snapshot view of investigative practices at a particular time and in a certain locale. In reality the practices we are interested in constitute a moving target in the sense that they are subject to historical change, partly because of conflicts within the scientific community, and partly because they are embedded in a wider context of historical change (the outermost of our three circles). Thus, we will have to extend our analysis over several decades to get a sense of the trend of historical development of investigative practices. Because the sheer volume of research output makes it practically impossible to analyze every relevant published article, it will be necessary to sample publications at an interval of several years.

We will also have to limit the scope of our inquiry to one part of the history of modern psychology. There can be no question that that part will have to be the earlier rather than the later period in the development of the discipline. Not only would it be difficult to achieve an understanding of later historical developments without a knowledge of the preceding period, but there would be a real danger of missing some of the most significant issues. By the middle of the twentieth century psychological research practices had settled into a rather rigid mold that allowed for relatively little variation.[35] But during the first half-century of its existence the investigatory practices of the new discipline were still in flux and diverged widely from each other. The roots of later rigidity lie in this earlier period, so that an analysis of what happened in this period can contribute to an understanding of the more familiar patterns that followed.

These temporal limitations allow a more comprehensive coverage of the research literature in the field than would otherwise be practicable. Such coverage is important because of the systemic level of the projected analysis. The kinds of processes we will be dealing with here are essentially collective processes that characterize the common practices of many individuals. Individual contributions are of interest only insofar as they represent a significant general trend. Thus, the role of specific individuals becomes relevant to our analysis primarily at the beginning of our period when modern empirical psychology first emerges as an identified field of activity for a distinct group of practitioners. Very soon, however, certain traditions of investigative practice form within the new field and subsequent developments are due to collective responses to conflicts among rival traditions and to pressures from outside. Although we need to understand the contributions of a few particularly influential pioneers at the point of emergence of the new discipline, the formation of distinct research communities soon makes it more appropriate to adopt a different level of analysis.[36] In the main we will be interested in identifying the pattern of variation, both synchronic and diachronic, that characterized the investigative practices of the discipline as a whole, rather than in assessing the role played by specific individuals.

2

Historical roots of the psychological laboratory

The birth date of modern psychology is usually placed toward the end of 1879 when Wilhelm Wundt designated some space at the University of Leipzig to be used for the conduct of psychological experiments.[1] Of course, the date is arbitrary, as are all such birth dates for disciplines. This arbitrariness arises less from the rival claims of other locations or individuals than from the obvious fact that the birth of a discipline is not a singular event but a complex process extending over a considerable period of time. In the case of psychology the relevant period extends both before and after the magic date.

When psychology became an autonomous field of research it did not invent its concepts and problems out of the blue but took them over from already existing fields like philosophy and physiology. Similarly, the practical activities that came to be identified as methods of psychological research were anything but completely new inventions. They were more in the nature of adaptations of already existing practices to a somewhat different context. Those who laid the foundations of a new psychological research tradition were constrained at every step by the traditions of investigative practice with which their general cultural experience or their personal training had made them familiar. The best they could do was to modify these practices, sometimes in quite minor ways, to suit the new goals they had in mind.

When Wundt proceeded to give practical shape to his program for a new kind of psychology, he based himself on three traditions of investigative practice, two of them explicitly, the other implicitly. Explicitly, he based himself first on the specific form of experimental practice that had recently grown up in physiology. This provided him with certain material techniques and a certain way of asking research questions. Second, he proposed to apply these procedures to a different object of investigation than the objects

17

addressed by physiological experimentation. The object he proposed was
the private individual consciousness. Historically the notion that this could
be a special object of investigation was closely linked with the investigative
practice of introspection. Wundt rejected introspection in its traditional
form, but he did accept its object. Hence in the present chapter we will
examine the historical roots of the notion that the "inner" world of private
experience could be methodically explored. Wundt's own adaptation of
the introspective method will be discussed in chapter 3, in the context of
the controversy that erupted around this topic a generation after the establishment of his laboratory.

Although it received far less discursive attention than introspection or
experiment, a third element in Wundt's investigative practice was absolutely crucial to the whole enterprise – the social organization of psychological experimentation. This aspect provides the strongest grounds for
locating the beginnings of experimental psychology in Wundt's laboratory,
for it was here that scientific psychology was first practiced as the organized
and self-conscious activity of a community of investigators. Moreover, this
community spawned others similar to itself, thus initiating a specific tradition of investigative practice. This development, more than anything else,
established modern psychology as a distinct field of inquiry that was the
preserve of an identified research community. Wundt established the first
community of experimental psychologists simply by adopting the prevailing
German university link between teaching and research in the form of an
institute where students could pursue their research. This resulted in the
emergence of certain social patterns of psychological experimentation,
which quickly became traditional. That development will be examined in
the final section of this chapter.

Introspection and its object before Wundt

Although introspection has not played a large role in the development of
psychology during most of the twentieth century, it was a very important
topic of methodological discussion at the time of the emergence of modern
psychology and for several decades thereafter. Far from being the time-
hallowed method of prescientific psychology, as it was often presented by
behaviorist crusaders, introspection is actually a historically recent inven-
tion. Its roots may go back to certain tendencies in Protestant theology
that advocated a methodical self-examination of one's conscience. (Such
a practice would take over some of the functions of the public confessional.)
In English the term does not appear to have been used to refer to a
systematic and ethically neutral practice of self-observation until well into
the nineteenth century, and it certainly was not in general use until the
second half of that century.

Of course, philosophers before that time appealed to the subjective self-awareness of their readers, just as they appealed to their readers' observation of the actions of others, to their sense of logic, and to their religious and ethical beliefs. But these appeals were not regarded by them, nor can they be regarded by us, as constituting a methodology.[2] Philosophers' appeals to common experiences of subjective self-awareness were no more based on any specific methodology of introspection than their appeals to common experiences of social life were based on any specifically sociological methodology.

One must distinguish between our common everyday fund of experiences and those deliberate and systematic arrangements for the methodical production of certain kinds of experiences that we call methodologies. In the latter sense there certainly was no introspection before the nineteenth century.[3] The lexical change already noted appears to have reflected a change in the thing referred to, a new awareness of the methodological aspects of self-observation. In the English-language literature that awareness appears to have come about in response to certain Continental criticisms of the traditional philosophy of mind,[4] which went back to John Locke's distinction between two sources of knowledge – sensation and reflection. The former gave us knowledge of the external world, whereas the latter gave us knowledge of the operation of our own minds. Thus, to the philosophy of nature, based on the evidence of our senses, there corresponded a philosophy of mind based on the evidence of reflection. But this philosophical tradition made no distinction in principle between mere awareness of mental states and deliberate observation of such states.

That distinction was the product of Immanuel Kant's attempt at constructing a philosophical synthesis that would transcend the conflict of empiricism and rationalism. He accepted from Lockean empiricism the notion of a world of private experience, which becomes manifest to its possessor through the medium of an "inner sense," analogous to the outer senses that give us experience of the external world. But then he raised a question Locke had not raised: Can the experiences conveyed by the inner sense form the basis of a mental science just like the experiences of the outer senses form the basis of physical science? The answer was a decided no, because science, unlike everyday experience, involves a systematic ordering of sensory information in terms of a synthesis expressed in mathematical terms. The material provided by the inner sense was, however, resistant to mathematization, and so there could not be a science of mental life, or psychology.[5] Thus, although Kant did not deny an empirical basis to psychology, he considered that this was not enough to establish it as a science. For that a particular methodological element would have to be added, and in the case of mental life this addition would not work.

In addition to making a fundamental distinction between natural science and psychology, Kant made another, more important distinction for the fate of psychology in the nineteenth century – that between the very different domains of psychology and philosophy. In the Lockean tradition of mental philosophy that distinction did not exist, and therefore psychology lacked any clear identity as a special field of study with its own objects and goals. The Lockeans were primarily interested in providing an empiricist theory of knowledge, and in doing so they appealed to a particular theory of mental functioning. But they did not distinguish between the psychological problem of how particular mental contents were caused in the natural world and the logical or philosophical problem of how such mental contents could be said to constitute knowledge of the world.[6] As long as that distinction was not made, psychology and philosophy remained indissolubly fused and the question of a specifically psychological methodology could not arise. Empiricist philosophers could go on appealing to their readers' everyday self-awareness as they appealed to other aspects of their experience, without any feeling that this required any special methodology.

Kant, however, made a sharp distinction between mental life as it is present to subjective self-awareness and the general principles in terms of which that life is organized. The former is part of the empirical world, just like our perceptions of external reality, but the latter points beyond such an empirical world to a "transcendental ego" that is the source of the fundamental categories that characterize human experience in general. To use a somewhat simplified illustration, there is a huge difference between examining the factors involved in particular spatial perceptions and examining the implications of the fact that our perceptions are characterized by spatiality. With this distinction Kant clearly separated the domains of philosophy and psychology and thereby raised the question of psychology as a nonphilosophical empirical discipline.[7] Such a discipline would be empirical, because it would base itself on the evidence of the "inner sense" in the same way that the various natural sciences based themselves on the evidence of the external senses.

But Kant did not think that an empirical basis was sufficient to establish a field as a true science. The classificatory botany of his day, for example, was certainly empirical but Kant did not regard it as a real science. It simply collected specimens and categorized them in terms of the occurrence and cooccurrence of particular features. But it was unable to show why the features of the plant world took on the arrangement they did rather than some other equally possible arrangement. The Newtonian science of classical mechanics, on the other hand, was able to demonstrate mathematically why a system of moving physical bodies like the planets maintained a particular arrangement of its parts, given certain initial assumptions. What made mechanics a true science was the mathematization of its subject matter. Kant had reason to believe that the data of the "internal sense" would always resist mathematization, and so it followed that psychology would never be a true science.

Kant's contribution to the history of psychology was therefore ambiguous. Although he provided the field with a clear identity, separate from philosophy, he also awarded this field a rather low status. At best, it would be little more than a collection of contingent rules, a kind of natural history of the mind, lacking both the fundamental importance of philosophy and the rational consistency of science.[8] To become a science, its special method of introspection would have to yield to mathematical treatment in the way that the visual data of astronomy, for example, yielded to mathematical treatment. But this would not happen, and so the subject had no future as a science.

The nature of the change from Locke to Kant is worth examining a little more closely. Clearly, the change involved a set of interconnected distinctions, a process of differentiation in a certain sphere of intellectual work. Kant's writings expressed a new level of interest in the organization of intellectual work. Not only was there the usual modern interest in the knowledge acquisition process as such, but there was a recognition of the fact that there are different kinds of knowledge based on different practices. It was now accepted that intellectual inquiry divided into different fields, each of which was constituted by a trio of interdependent factors: a certain object of investigation, a particular type of question or problem characteristic of the field, and a specific methodology by means of which these questions are pursued. Although the definition of the object of investigation limits the problems that can be formulated with respect to it and suggests certain methods as appropriate for their investigation, the adoption of these methods will continually re-create the objects that they presuppose and constrain the questions that can be legitimately asked of them. Thus, the advance of introspection from an occasional use of common experiences to a self-consciously employed *method* of investigation would not have been possible without the previous definition of a part of experience as *inner* experience, but of course the systematic practice of introspection helps to create just such a world.

It is sometimes said that modern psychology owes its origin to the post-Renaissance dichotomy between the extended mechanical world of matter and the totally different world of the mind,[9] but this is only a half-truth. That dichotomy, which is also the point of departure for the Lockean philosophy, certainly created a potential object of specialized investigation, the inner world of the isolated individual mind. But in the philosophy of Descartes and Locke it remained a *potential* object of technical scrutiny. What these philosophies did was to propose a certain way of talking about human experience, a system of terms implying certain distinctions and divisions in our experience of life.[10] This was an indispensible first step in the construction of a new field of study, but it was only a first step. Within a language community that tagged certain experiences as "inner" and treated them as fundamentally distinct, the potential for a specialized study of such experiences existed. But for this potential to be translated into actuality a further development was necessary.

This development involved the elaboration of a certain organization of intellectual work that went beyond the creation of new language communities. It was an organization manifested in the growth of the sciences during the eighteenth century. As knowledge of the natural world accumulated, scientific inquiry gradually became more differentiated and more self-conscious. The question of the relationship between science, or natural philosophy, and the more traditional concerns of philosophy began to be of interest, particularly in Germany where scientific activity was relatively more likely to occupy members of university philosophy faculties. The result was a sharpened emphasis on the methodological element in the constitution of the various fields of study. Among other things Kant was clearly concerned with the professional role of the philosopher in an intellectual situation that was being rapidly changed by the prestige and the spread of natural knowledge. A major effect of some of his writings was to give to philosophy a powerful role in the adjudication of divergent knowledge claims.[11] With Kant there began a tradition that was to last for well over a century, where the academic philosopher assigned relative status and intellectual limits to the various fields of empirical study. The grounds on which he did this were largely methodological, and this produced a heightened concern with such matters among those who were forced to justify themselves before this court.

The result was that in the nineteenth-century German universities the question of psychology as a field of study became problematized in methodological terms. Three issues dominated this debate: introspection, mathematization, and experimentation. The first two went back to Kant, the third became important after the middle of the nineteenth century. Friedrich Herbart, Kant's successor at the University of Konigsberg, engaged in a monumental exercise to show that the mathematization of psychology was possible in principle,[12] and Herbartian psychology remained an influential school until the end of the century.

Introspection remained a contentious issue. It will be remembered that for Kant the examination of the individual consciousness produced only limited results in both a scientific and a philosophical context. Its philosophical limitations arose out of the limitations of all empirical knowledge. The evidence of the "inner sense" was just as much subject to these limitations as the evidence of the external senses. In both cases the knowledge conveyed was knowledge of phenomena, of appearances, while the real world remained hidden from view. Introspection was limited to the phenomenal self: "I know myself by inner experience only as I *appear* to myself."[13] The true basis of our mental life, however, the subject of pure apperception, cannot be examined by relying on "inner experience."

The influence of the Hegelian school in post-Napoleonic Germany only served to confirm the low evaluation of individual self-observation. With *Geist* (mind or spirit) conceived as an objective category embodied in cultural formations and manifesting itself in philosophical abstractions, the role left to the observation of the individual human mind was small indeed.

In general, nineteenth-century philosophers divided into two camps on the subject of introspection, depending on which aspect of Kant's ambiguous legacy they chose to emphasize. Some, like Hegel and Kant himself, devalued the philosophical significance of the evidence from inner experience, others were encouraged by the possibility of developing a new empirical discipline on this basis.[14] The critical factor in this divergence of viewpoints was always the degree of individualism that characterized each philosophical position. According to one tradition, represented in its purest form by the British empiricists, philosophical speculation had to be grounded in the experience of individual minds. According to another tradition, represented by German Idealism, philosophy was grounded in supraindividual principles of order.[15] The latter group rejected introspection, blaming the superficiality and unreliability of its results; the former group accepted introspection as a necessary basis of philosophy and psychology. The second group made a clear division between philosophy and psychology, whereas the empiricists tended to blur the distinction between psychological and philosophical issues. Where the self-conscious individual was regarded as the centerpiece around which the world had to be arranged, introspection was held in high regard as a method;[16] where the order of the world took precedence over the individual, the examination of the individual consciousness was regarded with suspicion or condescension.

As the order of the world always included the social order as an important, and often as the all-important, component, conceptions of the role of introspection were not unconnected with considerations of political philosophy. During most of the nineteenth century a positive attitude to introspection tended to go with a philosophy of liberal individualism, while negative attitudes were more likely to be found among those who stressed the priority of collective interests or institutional requirements.[17]

Even an apparently private technique like introspection turns out to be very much a form of social action. It achieves this on two levels. On an immediate, more technical level introspection becomes social performance insofar as its use as a methodology requires the employment of language for the categorization and communication of experience.[18] But here we are not directly concerned with this inescapable fact but with the consequences of introspection's historical status as an avowed method of collective inquiry leading to true knowledge. To discuss introspection as a *method* means discussing it as a *collective* project, and that means putting it in the context of cognitive goals and interests that are assumed to have a shared, social character. It is not introspection as an idiosyncratic habit that is of interest here, but introspection as a social project. As such, it occurs in particular historical contexts from which it derives its meaning. This is true even when it is practiced unreflectively, as long as there is a significant uniformity of practice and of context. This context endows introspection with a variety of social meanings and hence also determines the forms of actual introspective practice.

As we have seen, the emergence of the notion of introspection as a *method* was intimately linked to the emergence of psychology as a separate field of study with its own special subject matter. The belief in the existence of this subject matter, the private world of inner experience, was a precondition for any meaningful discussion of introspection as a method. And such discussions, as well as the actual practice of the method, tended to validate this belief. Thus, method and object of investigation mutually confirmed each other, a state of affairs we will meet again in connection with other methods. Different evaluations of the object led to different evaluations of the method. So there could be differences of opinion, not only about whether introspection was valuable but about what one was actually doing when engaging in this practice. Thus, introspection could take on different meanings, depending on how its object was seen as fitting into the scheme of things. But this in turn depended on general or specific social interests. In this respect too introspection was no different from other forms of social practice to which the status of "method" was assigned.

Experimental physiology and the investigation of functions

Discussions of experimentation in psychology have often suffered from a misplaced abstractness. Not uncommonly, they have referred to something called *the* experimental method, as though there was only one, and in the worst cases this singular category is identified with another, called *the* scientific method. This kind of abstraction can only be maintained by ignoring the history of science, which provides evidence for the existence of several kinds of experimentation.[19] Moreover, the different kinds of experimentation have quite varied historical roots. Therefore, an inquiry into the origins of psychological experimentation must be concerned with the specific historical forms of experimentation that played a role in the constitution of the new field. Although experiments in the physics of sound and optics undoubtedly affected early laboratory practice in psychology, the systematic and programmatic aspect of psychological experimentation was clearly not derived from physics but from physiology. Indeed, in the first exemplar of a psychology constructed on an experimental basis, the term "physiological psychology" functioned as a synonym for experimental psychology.[20] Wilhelm Wundt frequently talked about adopting the experimental method from physiology. Let us see what was being imported into psychology here.

When Wundt received his medical training, just after the middle of the nineteenth century, the experimental method had only barely established itself as the primary method of physiological investigation in Germany.[21] The older generation of physiologists represented by Johannes Müller (d. 1858) did not regard physiology as an *essentially* experimental science, although they might perform experiments from time to time. It was Müller's ablest students and their generation who changed that. But these men – Du Bois-Reymond, Ludwig, Helmholtz – were only about fifteen years older than Wundt himself, though by 1860 it was clear that their work was transforming the discipline of physiology. What the young Wundt obviously hoped was that this recent rather impressive success might serve as a model for the transformation of another field, namely psychology.[22]

Although his immediate predecessors in German physiology provided Wundt with his most patent source of inspiration, it was not this group that had invented the project of an experimental science of physiology. They were in fact following in the footsteps of French investigators earlier in the century, chief among them François Magendie (1783–1855). In post-Napoleonic Paris we find the first clearly defined and established community of investigators devoted to the pursuit of physiology as an experimental science. Although physiological experiments had been performed before, and others engaged in this practice, more than merely a difference of degree existed between those who sometimes resorted to forms of experimentation, embedded in other practices, and those for whom experimentation had become the *via regia* to physiological knowledge.[23] The elevation of experimentation to the defining practice of scientific physiology cannot be separated from fundamental changes in the conception of the field and of its characteristic objects of investigation.

Traditionally, physiology had not been considered a separate discipline but merely the junior partner of anatomy. Medical schools had common chairs of anatomy and physiology, with the latter usually understood as being in the subordinate position.[24] Associated with this institutional arrangement was a certain conception of the subject matter of physiology. Function was considered to be subordinate to structure; one started with the anatomical organ and looked for its specific function.[25] The body was a static hierarchy of organs, each with its characteristic function. Thus, questions of physiological function could only arise after the structures to which the functions belonged had been established anatomically. The unit of investigation was the visible anatomical element, and the preferred method was that of dissection.

The situation changed drastically with a development that was as much a change in the object of investigation as a change of method. Functions were no longer seen as the properties of visible anatomical units but as abstract objects of investigation that might involve several organs as well as invisible processes. Thus it was no longer the function of a specific organ, like the stomach, which was at issue, but the role of the stomach in a function like nutrition. Structures were now subordinated to general functions that involved the interplay of many organs and systems. A functional perspective was becoming dominant, and the most appropriate method for investigating the functions of living systems was that of experimental intervention.[26] The systematic use of experimentation ensured the predominance of the functional perspective, for the only kinds of questions that were asked and answered within this investigative framework were functional questions. Experimentation meant formulating one's inquiry in terms of the contribution of various factors to a particular functional effect. The choice of one's primary method entailed a certain conception of the object of investigation. For the science of physiology, experimental laboratory methods provided the "cognitive identity" that was needed to force its institutional separation from anatomy and to establish it firmly as an independent discipline, a development that was not finalized until well into the nineteenth century.[27]

Not only did the transformation of physiology extend forward beyond the first generation of systematic experimentalists; its roots can also be traced backward to an earlier generation. This becomes particularly clear in those aspects of physiological thought that impinge most closely on psychology. After the middle of the eighteenth century one can detect a reaction against an earlier tradition that had made an absolute separation between mental and physical causes and effects, between voluntary (mentally caused) and involuntary (physically caused) action. Fundamental issues had been decided on the basis of whether one was thought to be dealing with a metaphysical mind substance or a metaphysical body substance. But in the second half of the eighteenth century a number of medical investigators – notably Whytt, Unzer, Prochaska, and, to some extent, Haller – began to move away from this preoccupation and to formulate questions of animal motion in purely functional terms.[28] The antecedents of the actions of living beings were now defined in terms of their effects rather than in terms of their status as mental or physical entities, and with this we get the emergence of the modern conception of *stimulation*. One result of this was to legitimize the treatment of psychological topics in a physiological context.[29] From then on the subject matter of psychology was likely to be influenced as much by developments in physiology as by philosophical considerations.

In the long run, there were two major channels through which an experimental physiology of function influenced investigative practices in psychology. One of these involved the functional characteristics of reflex movement. Although this aspect had some early effects on psychological theorizing, it did not influence psychological practice until well into the twentieth century, and so will not be considered here.[30] A second channel involving sensory physiology became active at a much earlier date. When the new program of systematic experimentation became extended to sensory functions, some psychological implications could not be avoided. In these physiological experiments variations in sensory effects were investigated as a function of variations in the conditions of stimulation. But in terms of the prevailing metaphysic, sensory effects, at least in humans, were categorized as belonging to that private world of individual experience that was supposed to constitute the subject matter of psychology or mental philosophy. Thus, experiments in sensory physiology acquired a theoretical significance that often went considerably beyond the modest aims of their originators.[31]

When Wundt published the first textbook of the new experimental psychology, a good two-thirds of it consisted of an account of the physiology of the nervous system and of research in sensory physiology.[32] The effect of the latter was to put the discussion of psychological objects – that is, sensory experience – in a functional context. This was a necessary consequence of the decision to ground psychological discourse in the experimental practices characteristic of sensory physiology. The kinds of questions that could be addressed on this basis were questions about the functional dependence of aspects of sensory experience on conditions of stimulation, such as intensity, spatial location, and temporal duration. This was the form taken by most experimental psychological investigations[33] and also the major empirical content of texts in the new discipline for many years to come. If the tradition of mental philosophy, with its notion of introspection as a method, bequeathed to the new psychology the concept of an inner mental world as a potential object of study, the model of physiological experimentation left the new discipline no choice but to pursue this study in a functional framework. The investigation of sensation and perception was virtually the only area in which these two approaches could be effectively combined, and the greater the pressure to expand beyond these confines the greater the methodological difficulties that the new discipline faced. We will explore some of these difficulties in the next chapter, but first we will need to take a closer look at the social features of early psychological experimentation.

The social practice of psychological experimentation

So far we have treated experimentation as though it were an almost purely cognitive activity. This perspective, however, ignores the crucial point about modern scientific experimentation – what distinguishes it from alchemical experimentation, for example – and that is its public character. We are not dealing with a private activity, whose procedures are to be kept secret, but with a technique for producing a social consensus about "the facts." Scientific experiments are supposed to deal in phenomena that are in principle accessible to everyone and in procedures that should be replicable by all others with the necessary training and materials. A successful program of scientific experiments requires a *community* of investigators who are able to agree on the veracity of certain observations, either because they have actually shared in their replication, or because they are satisfied that they have sufficient information to produce the same conjunction of events if they wanted to take the trouble to do so. The effective functioning of such a community requires the acceptance of certain rules and conventions about the conditions under which phenomena are produced and witnessed, and about the manner in which such matters are to be publicly communicated. Experimentation is not just a matter of cognitive construction but also a matter of social arrangement.

The original modern experimental community was that which formed around the Royal Society in England just after the middle of the seventeenth century.[34] Subsequently, new communities of the same general type formed in other centers and by the nineteenth century such communities generally had a more specialized focus, like experimental physiology. With each extension of the basic model some specific variants of the rules and conventions governing the life of the scientific community had to be adopted because of local conditions or special features of the objects of investigation. But this did not affect the fundamental features that characterized all such communities. They remained devoted to the engineering of an internal social consensus on so-called matters of fact by means of certain techniques that were common to them all. These techniques converged on the production of a sharp separation between matters of factual observation, about which all would necessarily have to agree, and matters of theoretical interpretation, about which differences of opinion within the community were permissible. Among the techniques deployed by the experimental community, those that regulate public communications within the community and the social relations of members of the community are of particular interest here.

The critical step that Wundt took toward the formation of a new experimental discipline was not the publication of his textbook, significant though this was, but the setting aside of a special space, identified as a psychological laboratory, for the regular conduct of psychological experiments by his senior students and himself. This happened at the University of Leipzig about five years after the first appearance of his textbook, and it was soon followed by further steps toward the institutionalization of this new area of experimental practice.[35] In 1883 the Leipzig laboratory received official recognition and a budget as a scientific institute, and in the same year there appeared the first number of a new journal, *Philosophische Studien,* which regularly published experimental reports from this laboratory. Wundt was fortunate in finding himself in the right place at the right time. The University of Leipzig was second only to Berlin in size and endowment at a time when the whole German university system was expanding rapidly,[36] and the prestige of German science and of the experimental method were at their height. He quickly gathered in his laboratory a respectable number of students, many of whom proceeded to produce doctoral dissertations based on their experimental work.[37]

Rather than work in isolation, several students would be doing experimental work at any one time, sharing facilities and assisting each other in various ways. They saw themselves as working in the same field, they shared certain theoretical interests, and their various experimental projects frequently had a bearing on each other. In other words, these students, together with Wundt himself, formed a viable community of experimenters. As some graduated and left, others came to take their place, so that practices that had become traditional in the community not only continued within it but were transplanted to other centers where the newly qualified experimental psychologists sometimes attempted to establish laboratories on the Leipzig model.[38] In Germany an independent second laboratory emerged at the University of Göttingen, and in 1890 a second journal, the *Zeitschrift für Psychologie und Physiologie der Sinnesorgane*, made its appearance.

For this experimental community to thrive, as it obviously did, and to pursue its work with some sense of effectiveness, certain technologies were needed to regulate its internal life.[39] The material technology could be taken over virtually intact from experimental physiology. The literary conventions of communicating the procedures and results of experimental investigations could also be adopted from existing models, and it was only gradually that the special features of psychological investigation led to new devices in the textual presentation of experimental results. In the early years of the experimental psychological community, the regulation of the social relationships inherent in the practice of psychological research presented the most interesting problems.

The classical work in sensory physiology – that of E. H. Weber, Hering, Helmholtz, and, for that matter, Fechner – had essentially been carried out by single investigators with, at most, occasional assistance. They communicated their findings to their scientific community in monographs and papers of a form that had become fairly standard in Germany by the mid-nineteenth century. The question of communication only arose for them when their work came to be published. But Wundt, as we have noted, established his laboratory explicitly for the benefit of his students, so that they might have a place to do the experimental work, which, for many of them, would form the basis of their doctoral dissertations and their first scientific publications. In doing this he was simply following the general trend in Germany of translating the ideal linkage between teaching and research into specific organizational forms.[40] Increasingly, the new sciences were linking the production of knowledge to the training of their recruits at universities, institutes, or laboratories, and Wundt did not hesitate to adopt this pattern.[41] But this meant that the work coming out of his laboratory was essentially collaborative work in which several individuals worked on different aspects of the same problem or on related problems.[42]

This organizational aspect of the research work in the Leipzig laboratory made it possible to share the tedium that characterized many of the investigations, especially in psychophysics. It also made it easier to exploit the possibilities of the relatively more complex recording apparatus that was becoming available.[43] But such advantages were available only insofar as the collaboration of research workers was extended from the usual convergence *between* investigations to a systematic division of labor *within* each investigation.

This was an extremely important development with the most profound implications for the nature of psychological research. The division of labor that was spontaneously adopted in Wundt's laboratory was none other than the well-known division between the roles of "experimenter" and "subject" in psychological experiments. This division had no fundamental theoretical significance for the Leipzig group and was undertaken as an essentially practical response to the conditions of psychological research in Wundt's laboratory. As in other scientific experiments the observational data that constituted the experimental yield consisted mostly of readings of instruments (e.g., time measurements) and simple sensory judgments (e.g., on the relative brightness or size of two presented physical targets). But these targets were categorized as psychological stimuli, that is, as events whose status depended on an individual's conscious response. In other words, variations in the sensory target were of interest not because of what they might indicate about its own physical nature but because of what they might signify about the nature of the consciousness that responded to them.

But this change of the interpretive framework within which individual observations were placed entailed certain changes in the way in which individual investigators interacted with the apparatus and with other investigators who might be assisting with the experiment. The individual consciousness, being the object of investigation, had to be shielded from variable internal and external influences of unknown effect, which might distort the particular response that was of interest. So it seemed desirable to get *immediate* responses that allowed no time for reflection and to keep the responding individual in ignorance of the precise short-term variations in the stimulus conditions to which he was to respond. (This is not to be confused with being ignorant of the overall purpose of the experiment, which was definitely not thought desirable.)[44] But this made it increasingly difficult for individuals to experiment on themselves without assistance. The task of simultaneously manipulating the apparatus and playing the role of the possessor of a shielded private consciousness whose precise responses were the object of investigation was not easy, and was sometimes downright impossible. To share the burden, one made use of the availability of other members of the laboratory, or of obliging friends.[45] However, the interpretive framework within which the work was done ensured that this collaboration would take the form of a division of labor between those who manipulated the apparatus and those whose consciousness was under investigation.

The effect of these practical measures was certainly not foreseen and hardly noticed for several generations. What occurred in practice was the development of a fundamental difference between the social conditions of experimentation in the natural sciences and in psychology. In the natural sciences any division of labor within an experimental investigation was unconnected with the fundamental relationship of the investigator to the object of investigation. One person might take charge of the distillation process while another might handle the weighing of the residue, but this did not affect their relation to the chemical object in any fundamental way. However, in psychological experiments one person would function as the repository of the object of investigation, of the data source, while the other would merely act as experimental manipulator in the usual way. This meant that whenever this division of labor was adopted the outcome of the investigation was the product of a social interaction within a role system whose structure was intimately connected with the way in which the object of investigation had been defined. In the natural sciences any division of labor in the experimental situation could continue more or less on an ad hoc basis, but in psychological experimentation the division of labor between experimenters and experimental subjects rather quickly developed into a universally accepted structural feature of the psychological experiment as such.

Once the organized psychological laboratories were established, self-experimentation began to decline in importance until, quite soon, it contributed only a small fraction of published experimental reports. This appears to have been the result of the conjunction of two factors. One factor must be looked for in the general development of institutional forms for scientific work, which favored collaborative research in organized laboratory units. This was not peculiar to psychology, except that psychology was affected at the very beginning of its existence as an experimental science, so that it hardly knew any experimental traditions other than highly organized ones. The second factor was connected with psychology's peculiar subject matter, which was the individual private consciousness. This made it inevitable that whenever an experimental division of labor was adopted it would take the form of a division between the roles of experimental manipulator and experimental subject or data source. As a result, the product of psychological experimentation depended not only on the interaction of investigators who all had the same fundamental relation to the object of investigation, but also depended on a unique kind of interaction between the experimenter's role and the experimental subject's role.

The special social organization of the psychological experiment is reflected in certain linguistic terms that appear even in the earliest published reports of these experiments. When describing procedural aspects of experiments, the research articles had to refer not only to the apparatus but also to the human participants in the experimental situation. Generally, these human participants were referred to in terms of the activity they had been assigned by the experimental division of labor. So we get references to "the discriminator," "the associator," or "the reactor," depending on the particular activity that the experimental task required. Similarly, those functioning as what we would now call experimenters were referred to variously as "the manipulator," "the signaler," and "the reader" (*der Ablesende*), because of their function of manipulating the apparatus, signaling the experimental stimuli, or reading off the experimental results.[46] While the participants were also identified by name elsewhere in the publication – a point to which we will return in a later chapter – the procedural references in these early research articles clearly describe an interaction system, not among historical individuals, but among the occupants of specific roles within an experimental division of labor.

If one traces the way in which participants in psychological experiments are referred to in the early empirical papers of the Leipzig group, one can detect evidence of a rapid process of institutionalization, which manifests itself in two ways. First, the more concrete terms used to refer to experimental participants tend to drop out and to be replaced by more abstract terms. The relatively specific activities that the experimental division of labor had assigned to them – activities like signaling or associating, as already noted[47] – were soon replaced by labels that refer in a more general way to the function of the participant in psychological investigation. The favored terms that now emerge are "observer," "reactor" (*Reagent*), and "person under experiment" (*Versuchsperson*) on the one hand and "experimenter" on the other. This means that individuals performing very different specific activities in different experiments are now referred to by the same general term. In other words, linguistic usage begins to reflect the fact that all psychological experiments involve a system of distributing certain general functions among the participants, and that these general functions are not tied to any specific activity.

The second, related development of linguistic usage in the early experimental psychological literature involves a certain standardization of terminology. During the early years of the Leipzig laboratory there is no uniformity in the labels applied to the experimental participants. Even when two investigators are concerned with the same research topic, they might use different terms to refer to their experimental subjects. For instance, two Leipzig experimenters investigated the "time sense" within a couple of years of each other, yet the one refers to his subjects as "reactors," whereas the other uses the term "observers."[48] This lack of semantic uniformity is also reflected in the fact that some investigators use two or three terms interchangeably within the same experimental report to refer to their experimental subjects.[49] But gradually usage becomes more standardized, even though complete standardization was not achieved till much later, as we will see in chapter 6. Before the end of the nineteenth century, however, most investigators settle into the consistent use of either of two terms to refer to the human data source in psychological experiments, the terms being "subject" (*Versuchsperson*) and "observer" (*Beobachter*).

The social process reflected in these developments clearly involved the institutionalization of psychological experimentation. In the 1880s the Leipzig laboratory evolved a distinct pattern of practice that defined for its participants what a psychological experiment looked like. Many of those who spent time at Wundt's laboratory subsequently used its pattern of experimental practice as a model to be emulated when they attempted to establish their own laboratories elsewhere. This model character of the Leipzig laboratory extended not only to the inventory of apparatus[50] and the kinds of research problems that defined a psychological laboratory but also involved the social arrangements required by psychological experimentation. Thus there grew up a certain tradition of how the social organization of work in psychological laboratories was to be structured. It was a tradition that depended on a certain definition of the object of psychological investigation, on a specific heritage of experimentation in sensory physiology, and on some local historical conditions that gave a particular form to the social conditions of psychological experimentation. But the effectiveness of this tradition would extend only as far as the historical conjunction of these three factors. Changes in any of these factors would produce a different model of psychological experimentation. This development will be examined in the next two chapters.

3

Divergence of investigative practice: The repudiation of Wundt

Wundt's limitations on introspection, experiment, and scientific psychology

Wundt's historical role was in some ways akin to the fate of the sorcerer's apprentice. He successfully mobilized some very effective practical techniques in the service of certain limited goals, but then found that these techniques turned into forces that had passed completely out of his control and were about to destroy the very framework within which he had put them to work. This was true of the link between introspection and experiment, which he advocated, and of the form of experimental psychology that he put into practice. Wundt, as has been observed, had many students but no genuine disciples. If he was the father of a discipline, he was a father whose fate is more reminiscent of the myth of *Totem and Taboo* than the myth of Chronos. Virtually everything that happened in modern psychology was a repudiation of Wundt, explicitly or implicitly. To understand this process, we must try to understand its starting point, namely, Wundt's conception of psychological practice.

Wundt's theory and practice of *introspection* diverged quite sharply from that of many of his students. On this issue (as on several others) he never emerged from the shadow of Kant, which meant that he basically accepted the object of knowledge to which the method of introspection corresponded but denigrated the method itself. He never doubted that the private consciousness was the object that psychology had to study, but he agreed with Kant and later critics like Comte and Lange that introspection was not the instrument that would transform this object into a scientific object. In fact, he went so far as to ridicule the introspectionist, likening him to a Baron Münchhausen attempting to pull himself out of the bog by his own pigtail.[1] Yet Wundt thought he had found a way to make consciousness available to scientific inspection in spite of these objections. How did he propose to achieve this?

At the basis of Wundt's proposal there is a distinction, also made by Brentano at about the same time,[2] between actual introspection (*Selbst-beobachtung*) and internal perception (*innere Wahrnehmung*).[3] It is a distinction between simply perceiving subjective events, being aware of them, and observing them in some methodical way. (As we saw in the previous chapter, some such distinction is in fact entailed by Kant's approach to the topic.) Now, as Wundt saw it, the problem for a scientific psychology of mind was the creation of conditions under which internal perception could be transformed into something like scientific observation. It was not enough to turn one's gaze inward in an attempt to give a systematic account of one's experience, for then one would not actually be observing ongoing events but more likely reflecting on what one thought one's experience had been. This was what the introspectionists of ill repute had done, and it was no basis on which to build a genuine science of mental life. The way out, for Wundt, was the manipulation of the conditions of internal perception so that they approximated the conditions of external perception. This manipulation was accomplished in the psychological experiment.[4]

In the laboratory, observation and report could follow immediately on the original perception, without time for self-conscious reflection. The conditions of internal perception would be closer to those for external perception where one reported on phenomena as they were happening. The conditions of psychological observation would thus approximate those of ordinary scientific observation. Second, according to Wundt, the laboratory made it possible to replicate specific experiences at will in order to observe them. We can make use of the fact that under certain circumstances identical external stimuli produce the same or very similar subjective experiences in order to scrutinize these experiences as often as we like and to be ready for them when they occur. This possibility of replicating observations again moves internal observation closer to normal scientific observation. The general idea was that internal perception could yield acceptable data for science only insofar as experimental conditions permitted a replication of inner experience at will. This rested on the assumption that there were external stimuli whose repeated presentation would reliably produce identical or near-identical perceptions. Wundt sought to achieve these conditions by limiting himself to relatively simple sensory stimuli and severely restricting the kinds of judgments required of experimental observers. In practice, so-called introspective reports from his laboratory were largely limited to judgments of size, intensity, and duration of physical stimuli, supplemented at times by judgments of their simultaneity and succession.[5]

A significant number of studies emanating from the Leipzig laboratory did not even contain this much "introspection." In these the experimental data simply consisted of time measurements, as in the reaction-time studies to which Wundt attached much theoretical importance. Another set of studies used response measures that depend on the activity of the autonomic nervous system. A kind of introspection would sometimes be called upon to check on the effectiveness of experimental manipulations (e.g., fluctuations in the level of attention), but purely introspective data were not recognized as a basis for knowledge claims in the Leipzig laboratory. Wundt would have been horrified to find himself classified as an introspectionist. Nevertheless, he held fast to the traditional notion that the proper object of investigation for the field of psychology was the human mind. This tension between method and subject matter lies at the root of psychological research practice in its earliest period.

Wundt had his own set of methodological prescriptions for dealing with this tension. They led him to impose very severe limitations on the scope of the experimental method in psychology. To understand the early history of scientific practice in psychology, one has to appreciate the Janus-like face presented by its most important figure. The same Wundt whose laboratory functioned as the model and inspiration for numerous imitators was also the source of a mounting stream of restrictions on the use of the experimental method in psychology.

One set of limitations was that which affected the *range* of topics and problems to which the experimental method could be validly applied. This range was defined by the reliable coordination of specific external conditions and specific internal conditions open to internal perception. Effectively, this kind of coordination could only be achieved in experiments on sensation and perception. Wundt had hoped that a similar coordination might also be achieved on the response side, initially involving time measurements[6] and later also measurements based on autonomic arousal. But essentially this hope proved to be unjustified, which may have contributed to his tendency to become less and less sanguine about the prospects for the experimental method as time went on. At all times Wundt explicitly excluded certain parts of psychology from the possibility of effective experimental investigation.[7] Specifically, these were the processes of thought and all but the simplest affective processes. Here there seemed no hope of achieving that direct and reliable coordination of external conditions and internal perception that an effective psychological experiment presupposed.

In the course of his long career Wundt at times expressed himself with varying degrees of optimism or pessimism as to whether certain areas of psychology were likely to succumb to the experimental method. Such opinions, however, did not involve any change in basic principles. What he always accepts, as an inescapable corollary of his fundamental principles of methodology, is the notion that the area of psychology cannot be coextensive with the area of experimental psychology. The appropriateness of experiment for the solution of various types of psychological problems is a matter of degree. At one extreme are problems for which the experimental method provides an excellent source of valid data, at the other extreme are problems that are quite unsuitable for experimental investigation. Wundt consistently assigns problems in the areas of sensation and perception to the top end of this quasi scale and problems in the areas of thinking, affect, voluntary activity, and social psychology to the bottom end. In between there are areas, such as memory, imagery, and attention, where the experimental method is partially appropriate. Precisely where the line is to be drawn at any time depends partly on technical developments and partly on the optimism of the investigator, and hence the line is historically variable. But such decisions do not affect the fundamental frame of reference within which they are made.

Even in his period of youthful enthusiasm for the experimental method Wundt recognized the need for another, nonexperimental type of psychology. This was *Völkerpsychologie,* an untranslatable term referring to a kind of social psychology based on the historical, ethnographic, and comparative analysis of human cultural products, especially language, myth, and custom.[8] For all his interest in experimentation Wundt was sufficiently immersed in the German tradition of objectifying *Geist* to reject the notion that a study of the isolated individual mind could exhaust the subject matter of psychology. But as his conception of psychological experimentation made precisely the isolated individual consciousness its object of study, the only recourse was to supplement it by a different kind of investigation directed at mind in its objective manifestations. These investigations occupied Wundt increasingly in his later years and resulted in the series of volumes that he published under the general title *Völkerpsychologie* from 1900 onward. He now expressed himself to the effect that this was the more important part of psychology, which was destined to eclipse experimental psychology.[9]

Although Wundt's formulation of the precise relationship between experimental psychology and its complement underwent several changes in the course of his long life,[10] he stuck to three fundamental points: First, that experimental psychology could never be more than a part of the science of psychology as a whole; second, that it needed to be supplemented by a branch of psychological studies that was devoted to the investigation of human mental processes in their social aspects; and third, that this latter type of study was able to make use of information that was no less objective than the data of experimental psychology.

But Wundt not only imposed severe limitations on the *range* of questions that the experimental part of psychology could handle, he also held a general theory of scientific method that tended to circumscribe the role of experimental evidence in the scientific enterprise as a whole. His view of experimental science was decidedly antiinductivist.[11] The aim of scientific explanation was the demonstration of the logical coherence of the world by revealing its underlying causal relations. One performed experiments in pursuit of this aim, not to accumulate data from which generalizations might follow.

The world of physical science, according to Wundt, had been created by ignoring "the subjective elements of perception." This made possible the universe of physical causality but also implied that if the neglected elements were to be scientifically investigated there would emerge a science that would "supplement physics in the investigation of the total content of experience." This science was psychology.[12] Whereas physical science would frame its explanations in terms of physical causality, psychology, by analogy, would be concerned with something Wundt called *psychic causality*. The specific principles of psychic causality that Wundt attempted to formulate[13] always remained a bit vague, but this does not affect the fundamental role this concept played in his approach to psychological investigation. Whatever their precise form, Wundt believed that there were psychological determinants at work in experience and that it was the task of psychological experiments to explore their mode of operation. Psychological investigation had to be experimental, rather than merely observational, because what it sought to illuminate were not static structures but directed causal processes.[14] Wundt was, however, insistent that the laws of psychic causality would be qualitative in nature, and this limited the functionalism of his experimental approach.

It is clear that Wundt's practice involved a certain leap from the experimental to the theoretical level, and this is closely related to the tension between method and object of investigation, which has already been noted. For psychological experiments to be relevant to psychic causality, it was not necessary for their data to consist of introspective reports. It was necessary, however, that they be conducted with individuals who were in a position to provide such reports. For the aim of such investigations was the exploration of psychic causality, and this, according to Wundt, only operated within fully conscious human experience. If one worked with animals, or with individuals in pathological states of consciousness, one had given up psychology's proper object of investigation. Thus, although Wundt hardly qualifies as an introspectionist on the level of experimental practice, he always remained mentalistic on the level of theoretical explanation. Without a mentalistic object psychology would not have a distinct subject matter of its own, although without an experimental practice based on physical manipulation and measurement it would have nothing trustworthy to say about its subject matter.

This profound duality in Wundt's investigative practice was undoubtedly a reflection of his failure to cut himself off from his divergent physiological and philosophical roots. He hardly altered the physiological techniques he had inherited, and neither did he propose any fundamental change in the object of investigation, which he had taken over from certain trends in philosophy. But he made no significant changes because it is abundantly clear he had no desire to make such changes. Wundt had no interest in two goals that were to become crucial for his successors and that were to lead to the transformation of modern psychology: He had no interest in the possible practical applications of psychology and he had no interest in converting psychology into an independent discipline without ties to philosophy.

Wundt belonged to a generation of German academics for whom the refusal of practical social involvement outside the university was effectively a condition of academic freedom.[15] At the same time he was quite content with the existing division of academic labor that allocated psychology to philosophy, a subject to which he had made his own direct contributions.[16] He saw his work in psychology as essentially another contribution to philosophy, although one that he hoped would have a major impact on that subject.[17] What Wundt sought to achieve was a rejuvenation of philosophical inquiry by new means, not the constitution of a completely new discipline.[18] For this the traditional object of psychological inquiry had to be preserved even as the means for pursuing that inquiry had to be radically changed.

A change in psychology's disciplinary project

Wundt's program for the reform of philosophy and the *Geisteswissenschaften* through the new psychology remained one man's idiosyncratic vision. It could please neither the more traditionally minded philosophers nor those modernists who considered philosophical goals irrelevant to the pursuit of knowledge through specialized scientific inquiry. In particular, Wundt's restrictions on introspection and experiment worked against the plausibility of his wider claims. These restrictions meant that the yield of his experimental program was actually rather limited. One could accept this with good grace and restrict the role of the experimental psychologist to that of a competent craftsman in a small number of specialized areas that lacked major intellectual or practical significance. Some, like G. E. Müller at Göttingen,[19] adopted this course, but in the intensely competitive academic world of the nineteenth century this was not a generally useful prescription for either personal or disciplinary success. It could work on a large scale only where a discipline already possessed an established intellectual and institutional framework, but of course this was precisely what was lacking in the case of psychology.

Wundt's legacy to his students contained both positive and negative elements. On the positive side there were the institutionalized arrangements for conducting a psychological laboratory. But on a more strategic level Wundt's vision for the new psychology was of less help to his students who faced a different world and had different ambitions. The root difference involved what is most appropriately called a "disciplinary project." By this is meant the practitioners' conception of how their discipline fits into the existing institutionalized structure of knowledge-generating domains and the nature of their discipline's contribution to public knowledge and social practice.[20] Such a project may be an expression of already established patterns or it may be more on the level of the desirable, but in any case it provides the context that gives a broader meaning and direction to specific types of investigation.

Wundt's disciplinary project for psychology did not provide for the severing of its links with philosophy and its development as an independent discipline. On the contrary, when some of the younger generation of experimental psychologists made proposals in this direction, Wundt opposed them vehemently.[21] His methodological prescriptions were entirely consonant with his project for a psychology that was to retain intimate ties with philosophy and the *Geisteswissenschaften* – psychology was to be only partly an experimental science.

This project, however, proved to be out of step with the development of the social conditions on which it depended for its success. Wundt's own appointment at Leipzig occurred at a time when the psychologizing of philosophy constituted a still-rising tide in Germany and aroused relatively little opposition.[22] It is understandable that he then took these conditions for granted and based his project for the future of psychology on their continuation. That would have meant that a significant proportion of German university chairs in philosophy would gradually be occupied by psychologists. Such a process did in fact begin to get under way, but in due course it evoked a most determined reaction among the philosophical establishment who did not in the least share Wundt's views on the relevance of scientific psychology for their own discipline.[23] His plans for the psychological rejuvenation of philosophy and the *Geisteswissenschaften* ran directly counter to the major internal developments in those fields, so that experimental psychology in Germany found itself rejected by the forces that were to have provided it with an institutional home base.[24] It was increasingly *forced* to make its own way as an independent discipline.

By contrast, in the United States, the advantages of independent disciplinary development were clear from the start, given a university structure that favored specialization and the absence (except on a local basis) of anything that could be compared with the entrenched German philosophical establishment. In Germany there had been a long period during which the sectarian religious control of university teaching had been replaced by the authority of philosophical systems, and in the late nineteenth century the institutional relics of this period were still strong. In American universities scientific – or, more properly, scientistic – dogma was apt to take over directly from religious dogma. Moreover, ultimate control of university appointments and professional opportunities was vested in businessmen and their appointees, in politicians and in men engaged in practical professional activity, rather than in state officials whose vision had been shaped by their own philosophical education. Thus, the representatives of a new discipline like experimental psychology had to legitimize their project before a very different tribunal in the two cases.[25] Although the project of a scientific psychology emerged in both settings, it was much more clearly linked to the project of an *autonomous* discipline in America than in Germany.

Contrary to Wundt's conception, the new psychology did not prosper through the links with philosophy, linguistics, history, and anthropology that he had tried to forge. Instead, it shifted its weight to its other foot, as it were, and based its claims for recognition entirely on its affiliation with the natural sciences. In this it was greatly encouraged by a swelling tide of scientism during the closing years of the nineteenth century. The triumphal progress of the natural sciences helped to promote the belief that their methods were the *only* methods for securing useful and reliable knowledge about anything. Insights and explanations that could not be tied directly to these methods were empty speculations unworthy of serious attention. The split between the natural sciences and the social and humanistic disciplines had finally become an unbridgeable chasm and psychology was caught in the middle, forced to one side or the other. What Wundt had tried to keep together, however tenuously, now broke apart irrevocably. Unsurprisingly, most of those who had come to him primarily because of his laboratory, including virtually all his American students, ended up on the side of natural science and experiment, consigning the rest of Wundt's psychology to the graveyard of metaphysics.[26]

There were two major themes in this surge of scientism. The one was technical-utilitarian, involving an assimilation of conceptions of truth to conceptions of usefulness and a consequent emphasis on practically applicable knowledge. Scientific knowledge was basically useful knowledge and thus superior to idle speculation. The second theme derived the superiority of scientific knowledge more from its close grounding in directly observed facts and its avoidance of airy metaphysical fancies that lacked this grounding.[27] Both themes were of course normative in the sense that they did not simply register what science was but prescribed what good or real science ought to be. It ought to be useful, and it ought to avoid theories that went beyond a summary of what had been observed. Both tendencies had their bogeyman. For the first, he was the ivory-tower intellectual; for the second, he was the philosopher spinning his metaphysical web.

These themes were obviously quite consonant with each other and they converged in their effect. But they appealed to different individuals in different degrees, depending on their backgrounds and life circumstances. Thus we find some of the first generation of post-Wundtian psychologists putting more emphasis on one or other of these aspects. There were the practicalists, like G. Stanley Hall, who were often far from meticulous in gathering observations and not going beyond them, and there were the experimentalists, like Titchener, who were sticklers for laboratory finesse, but who cared little for any practical applications of psychological knowledge. Nevertheless, both sides were united in their rejection of the Wundtian synthesis. Psychology was to be a specialized positive science, divorced from philosophy, but looking to the natural sciences for its theoretical and methodological models.

Leaving the practicalist orientation for later consideration, we must briefly note the impact of the experimentalist orientation on investigative practice. For neither practicalism nor experimentalism could help changing the practice of Wundtian experimental psychology. As we have seen, the latter's practice had been based upon a strict limitation on the proper topics for experimental psychological investigation. The rest of psychology was left to be elucidated by *Völkerpsychologie* and by theoretical excursions into the mode of operation of "psychic causality." Now, both of these aspects were rejected as "unscientific" or "metaphysical," and experimental observation was to constitute the whole of psychology. But the traditional mentalistic objects of psychological investigation were not given up. The only way of reconciling these investigatory objects with the new emphasis on experimental exclusiveness was by greatly extending the scope of experimental introspection.

Dead end: Systematic experimental introspection

Some of Wundt's most influential students, notably Külpe in Germany and Titchener in America, took this path and thereby inaugurated a rather odd episode in the history of psychology's investigative practice, the period of "systematic experimental introspection." It was a brief period, for the methods that characterized it were at the forefront of disciplinary advance for at most a decade, from about 1903 to about 1913. But it was a fateful period because it revealed many of the problems inherent in the Wundtian synthesis of rigorous experimentation and mentalistic objects. The ensuing crisis virtually eliminated the Wundtian program and favored the development of psychology's investigative practices along altogether different lines.

One problem that we face in trying to understand the new approach to introspection on the part of some of Wundt's most prominent students is historiographic in nature. The behaviorist turn of American psychology produced a certain foreshortening of perspective on everything that preceded it. Like all effective leaders of movements, J. B. Watson made good use of a simplified image of the opposition. In due course it became customary for the most diverse positions to be lumped together under the poorly understood category of "introspectionism." This had the effect of obscuring the very fundamental differences that existed between Wundt and his students, differences that were critical for the development of psychology as an independent discipline.

The change in the deployment of introspective methods certainly seemed very significant to those who helped to bring about the change. Here is how Titchener describes the situation in 1912:

> Those who remember the psychological laboratories of twenty years ago can hardly escape an occasional shock of contrast which, for the moment, throws into vivid relief the difference between the old order and the new. The experimenter of the early nineties trusted, first of all, in his instruments; chronoscope and kymograph and tachistoscope were – it is hardly an exaggeration to say – of more importance than the observer.... There were still vast reaches of mental life which experiment had not touched;... meanwhile, certain chapters of psychology were written rather in the light of "system" than by the aid of fact. Now twenty years after we have changed all that. The movement towards qualitative analysis has culminated in what is called, with a certain redundancy of expression, the method of "systematic experimental introspection."... A great change has taken place, intensively and extensively, in the conduct of the introspective method.[28]

What Titchener referred to as "extensive" changes included the extension of "systematic" introspective analysis to such areas as memory, thinking, and complex feelings, mostly topics that Wundt had explicitly designated as being unsuitable for the practice of experimental introspection. The Würzburg experiments were the most promising examples of this new fashion of intensive introspective analysis, but the vogue of qualitative introspection had many proponents and defenders who were not members of the Würzburg school.[29] In his relatively well-known attack on the methods of the Würzburgers, Wundt was of course expressing his opposition to the entire trend.[30] He reiterated that only when "the objects of introspection are directly tied to external physical objects or processes" did one have ideal conditions for experimental psychological investigation; all other circumstances compromise this ideal and need to be treated with great circumspection.

Among the "intensive" changes in the application of introspection that Titchener noted, two were particularly striking.[31] One involved a shift in the relative importance attached to objective and to introspective data within a particular series of experimental observations. In the more traditional kinds of experiments the essential data had been either completely objective, such as reaction times and errors of recall, or tied directly to measured variations in physical conditions, as in psychophysics and various experiments in perception. In terms of truly subjective data, the subject might at most be expected to give occasional reports on mental processes that accompanied his overt recorded responses. Systematic introspection changed this emphasis. Subjective reports were now required on a regular basis, usually for every experimental trial, and it was they, rather than more objective measures, that provided the essential data of the investigation. The studies of the Würzburg school, and those to which they gave rise,[32] provide the best-known illustration of this trend, although it is not limited to this group.[33] In the Würzburg studies the actual solutions of the experimental tasks become almost irrelevant to the real purpose of the experiment, which is to provide subjective data on the process of thought. To Wundt, of course, these were simply "pseudoexperiments." The more cautious Müller warned against the danger of these methods coming to be considered the norm in experimental psychology.[34]

Closely connected with the shift toward subjective reports is the second feature of the new introspectionism, the interest in qualitative description. Galton's studies of imagery[35] had to some extent anticipated this trend, but in the era of Fechner and Wundt they constituted a distinctly marginal area of psychological research. It was Binet who played a key role in challenging Wundtian introspection on its home ground. In 1903 he published a series of papers on investigations of the two-point threshold in which he had not limited himself to the customary introspective reports that went no further than noting whether one or two points had been felt upon application of the stimulus.[36] By questioning and through spontaneous reports from his subjects he accumulated a mass of qualitative data that showed that the conventional determination of the threshold represented a gross oversimplification of the subjective processes involved.[37] Binet's use of complex qualitative introspective material was quite similar in style to the methods of the Würzburg school, leading him to claim priority for himself.

The impulse that caused a significant number of experimental psychologists to turn toward "systematic" introspection at this time was the very opposite of "metaphysical." It was precisely because they wanted to replace the old-style speculation about matters like "psychic causality" by direct observation of mental events that they turned to qualitative introspective data. In doing this they were doing no more than following the time-honored basis of scientific experimentation, which provided for the creation of a distinct province of witnessed facts on which general consensus could be achieved.[38] Such a body of consensually validated facts would provide a much more satisfactory way of grounding disciplinary knowledge than this or that theoretical system. As is well known, the program of systematic introspection came to grief within just a few years. The reasons for this debacle are not without relevance for the history of psychology's investigative practices.

In the first place, we must distinguish between two kinds of criticism to which the practice of qualitative introspection was subjected. The first was primarily directed at the *object* of introspective practice, and the methodological aspect just followed as a consequence. Behaviorist criticisms of "introspection" were generally of this type. The fundamental objection was to making consciousness an object of scientific study. For J. B. Watson and most of his followers this rejection of psychology's traditional knowledge object was closely connected to their conviction that psychology had to be a practical science if it was to prosper as a discipline, and nothing practically useful had emerged or was likely to emerge from the analysis of consciousness.[39] The rejection of introspection as a method then followed as a matter of course, for the object on which it was to be practiced would cease to exist as an object for psychology. On the methodological side the behaviorist critique added nothing to what was already well known in the nineteenth century; the innovative aspect of this critique depended entirely on the radical change in psychology's investigative object.

In Europe there was little interest in a change of psychology's traditional knowledge object along behaviorist lines, but in many quarters there was dissatisfaction with the turn that experimental introspection had taken. This led to various kinds of criticism, some, like Wundt's, arguing for a return to the simpler certainties of a more traditional practice, others containing some new insights. These discussions did focus on the purely methodological aspects of introspection, sometimes entailing a change of investigative object as a consequence. In the present context we need consider only one line of criticism, which has a bearing on introspection as an investigative practice rather than merely as a technique.

From this point of view the critical development occurred when the problem of experimental introspection was recognized as essentially a problem of *communication* rather than a problem within the individual consciousness. The source of the difficulty did not lie in the privacy of the experiences that were observed but in the means available to establish a consensus that would transcend this privacy. In ordinary scientific observation the experiences that are described are private too – no two scientists see exactly the same phenomenon – but this problem is overcome by the use of special media of communication that make agreement on crucial aspects of the experience possible. These media are partly physical – relatively simple and artificially structured spatial patterns, for example – and partly linguistic. The role of the physical aspect had been recognized by Wundt and the more traditional experimentalists; now interest began to shift to the role of language. Was it possible to develop anything like a scientific language for describing the experiences evoked by "systematic" introspection?

The effective use of such a language would entail a certain attitude on the part of both participants in the experimental situation, experimenter and subject. The desired description would be of the kind that is used in talking about an object toward which one maintains a certain distance. Now, for the experimenter there is no special difficulty in talking about the subject's experience in this way. But for the subject there is, because it is his own experience from which he can only distance himself by adopting a special attitude, the introspective attitude. This is certainly possible, but what are the costs? For costs there must be, because something is lost between the original experience and the report to the experimenter. In the observational reports of natural science, much of the quality of the observer's experience is sacrificed too in the interests of reaching agreement on the essentials. But for traditional psychology the qualities of the original experience *were* the essentials. The achievement of the kind of consensus on the "facts" that science demanded would have entailed the abandonment of the object one was supposed to be investigating.

Proponents of "systematic experimental introspection" faced a dilemma. They could either insist on the use of the kind of distanced language that had been useful in achieving consensus in the natural sciences, or they could reject such limitations and try to remain close to the phenomena as experienced. In the first case they lost the real object of their scientific interest; in the second case they lost the possibility of achieving a scientific consensus on the "facts."

In practice the language of scientific description turned out to be the language of sensationism. By reducing lived experiences to reports on specific sensational components, systematic introspection could imitate the path taken by the observational reports of natural science. This policy was essentially the one advocated by Titchener and put into practice at his Cornell laboratory.[40] But Titchener was only able to argue in this way because he had reified the terms of his sensationist language and persuaded himself that the basis of conscious experience really was nothing more than an assembly of sensations. He now faced a war on two fronts, which could only end in total defeat. His fellow introspectionists refused to sacrifice their phenomenal objects at the altar of sensationist reductionism, and his more practically minded colleagues refused to be interested in a constructed object of investigation that had no relevance to everyday life.

"Systematic experimental introspection" was based on an attempt to extend the essentially public project of scientific witnessing to the most private areas of experience – complex and indefinite thoughts and feelings. But that would mean providing the same kind of unambiguous, analytical description of such experiences as one would give of a simple visual stimulus in a perception experiment. It was soon realized that if one did not, like Titchener, insist on such descriptions, the spontaneous reports of the subjects in these experiments were of a very different order. They were not at all like descriptions of an object regarded from a distance but more like a verbal expression that somehow belonged to the experience itself. Asked to report on their thoughts on receiving the experimental task, subjects would say things like: "Wait a minute, there's a trap here," or "well, really!" or "OK, here's another one of those Nietzschean paradoxes." But as von Aster pointed out in a brilliant paper,[41] the relation of the verbal "report" to the experience is quite different here than in the case where the subject reports, let us say, on the size of a presented rectangle or the length of a line. In the first examples the subject is giving *expression* to the Gestalt quality of his experience; in the second case he is *describing* specific elements of his experience. Note that he uses a metaphorical, expressive kind of language in the first case. He is not really describing something at a distance, he is giving public expression to a lived experience.

This insight led to the important distinction between *Beschreibung* ("description") and *Kundgabe*. The second term has no precise equivalent in English. It combines the connotations of "expression" and "communication." It refers to a process of giving public expression, a process of declaration rather than reporting. Its language, as von Aster noted, must necessarily be metaphorical and hence ambiguous, because we have no other way of giving verbal expression to an experience as a lived whole. But these features seem to render the language of *Kundgabe* unsuitable for the purposes of scientific communication. Only the neutral, analytic language of distanced description would do for that. But this language destroys what was supposed to be the object of investigation.[42]

"Systematic experimental introspection" therefore faced an impasse. If one rejected the Titchenerian destruction of actual lived experience, however, there were two possibilities that remained. One could try to develop a descriptive language that, while reasonably unambiguous, allowed the complex, wholistic qualities of experience to enter public scientific discourse. In due course, the Gestalt psychologists showed that this could be done, although it was done most successfully where the link to physical stimuli existed, in the study of external perception. In the area of thinking, it was always more difficult to carry through this program convincingly and consistently.

The second possibility opened up by a critical insight into the problems of "systematic experimental introspection" depended on an acceptance of the nature of *Kundgabe*. If it was true that in many psychologically interesting situations the subject's verbal statements were not really *about* his or her experience so much as an *expression of* that experience, then those verbal statements themselves should become the objects of psychological study. But they would be studied for what they were, a manifestation of a living personality, rather than as a source of information about a separate world of private consciousness. This of course would mean the abandonment by psychology of its traditional, investigative object, the world revealed by the "inner sense." In Europe this path led to Kurt Lewin's studies of motivated behavior as well as to some versions of clinical existentialist psychology. In America the whole development was of course short-circuited by the behaviorist turn. But one did not need to be a behaviorist to abandon psychology's traditional object of study.

These developments entailed a shift in the structure of the psychological experiment as a social situation. In the Wundtian experiment, the exper-imental subject was the scientific observer, and the experimenter was really a kind of experimental assistant. Only the person reacting to presentations of the stimulus material could observe the phenomena of interest and report on them. But if we start treating the experimental subject's report as *Kundgabe,* as an expression of rather than a descriptive report about his experience, then he is no longer in the role of a scientific observer. That role has now passed to the experimenter who records and then reports on what the subject said. It was noted at the time that this was happening in the Würzburg experiments.[43] The implications of this change were ex-tremely far reaching, for it demonstrated that the nature of the object of psychological investigation was linked to the social structure of the inves-tigative situation. A fundamental change in the one entailed a fundamental change in the other. As the object of psychological investigation changed, so the distribution of tasks in the investigative situation would have to change. The social structure of the Wundtian experiment could not survive the abandonment of its investigative object.

4

The social structure of psychological experimentation

How the Leipzig research community organized itself

In the last section of chapter 2 we took note of the fact that the establishment of psychological laboratories involved the institutionalization of certain social arrangements. Psychological experimentation became a collaborative effort dependent on a division of labor among individuals who carried out different functions in the experimental situation. More specifically, a broad distinction emerged between those who acted as the human source of psychological data, the experimental subjects, and those who manipulated the experimental conditions, the actual experimenters. Although this distinction was essentially a response to the practical exigencies of late nineteenth-century "brass instruments" experimentation and not the product of deep reflection, it rapidly became traditionalized. To carry out sophisticated experiments with relatively complex apparatus, one now had to assign the function of data source and the function of experimental manipulation to different people.

What was not noticed at the time, or indeed for almost a century afterward, was that this arrangement created a special kind of social system – one of psychological experimentation. The interaction between subjects and experimenters was regulated by a system of social constraints that set strict limits to what passed between them. Their communication in the experimental situation was governed by the roles they had assumed and was hedged around by taken-for-granted prescriptions and proscriptions.

However, the specific features of this social system were not necessarily fixed. The basic division of labor between experimenters and subjects still left much room for local variation. For instance, there was nothing in the practical requirements of psychological experimentation that dictated a *permanent* separation of experimenter and subject roles. The practical exigencies that had led to a separation of these roles did not preclude the possibility that the same person might take one or the other role on different occasions. All that experimental practice entailed was the convenience of separating these roles for a particular experimental session; it did not necessarily decree that people could not exchange these roles from session to session. However, other factors might enter the picture to convert the occasional separation of experimenter and subject roles into a permanent separation. Whether this in fact happened is a matter for historical investigation.

What kind of factor might be expected to have an effect on the permanence of the division of labor in experimental situations? One such factor is quite obvious. Psychological experiments are not conducted in a social vacuum.[1] The participants in such experiments are not social blanks but enter the experimental situation with an already established social identity. It is of course impossible to insulate the social situation of the experiment from the rest of social life to the point where the participants' social status outside the experiment has absolutely no bearing on their interaction within the experimental situation. Who gets to play experimenter and who subject may have something to do with the roles that individuals play outside the experimental situation. Similarly, the possibility of exchanging experimental roles may depend on the social identity of those who are occupying those roles.

Apart from the likely influence of social factors that are clearly external to the experimental situation, there is the question of how the distribution of roles in the face-to-face laboratory situation ties in with other activities that form a necessary part of the social practice of investigation. The direct interaction of experimenters and subjects in the laboratory is only one part of the research process. Activity in the laboratory is embedded in a penumbra of other activities, like theoretical conceptualization, data analysis, and writing a research report. The fact that in the contemporary psychology experiment these activities are assumed to form part of the experimenter role should not lead us to presuppose that this was always the case. The domain of what necessarily belonged to the experimenter role and what to the subject role has in fact been subject to historical variation. Let us therefore return to Wundt's Leipzig laboratory to see how subject and experimenter roles related to various other functions entailed by the social practice of psychological investigation.

When we peruse the research reports published in Wundt's house organ, the *Philosophische Studien,* we find that the person who had authored the published account of the experiment had not necessarily functioned as the experimenter. For instance, Mehner published a paper on an experiment in which he had functioned solely as the only subject while two other persons had functioned as experimenters at various times.[2] Moreover, functioning as a source of psychological data was considered incompatible with functioning as a source of theoretical conceptualization. During the first years of his laboratory, Wundt appeared regularly as a subject or data source in the experiments published by his students, although he also contributed much of the theory underlying these experiments.[3] Interestingly, Wundt does not seem to have taken the role of experimenter. This suggests that the role of psychological data source was considered to require more psychological sophistication than the role of experimenter. The role of experimental subject was quite compatible with Wundt's status as head of the laboratory, but the role of experimenter could be left to his students.

If the role of functioning as source of psychological data was quite compatible with the functions of a scientific investigator, there was clearly no reason to make a permanent distinction between experimenter and subject roles in the laboratory situation. Indeed, it was not uncommon for Wundt's students to alternate with one another as stimulus administrators and as sources of data, sometimes in the same experiment.[4] The roles of subject and experimenter were not rigidly segregated, so that the same person could occupy both at different times. Their differentiation was regarded as a matter of practical convenience, and most participants in the laboratory situation could play either or both roles equally well.

The participants in these experiments clearly saw themselves as engaged in a common enterprise, in which all the participants were regarded as collaborators, including the person who happened to be functioning as the experimental subject at any particular time. Characteristically, authors of experimental articles would sometimes refer to the experimental subjects as co-workers (*Mitarbeiter*), especially when giving their names as part of the published account of the investigation.[5] Moreover, functioning as subjects and experimenters for one another would usually be part of a relationship that extended beyond the experimental situation. Those who participated in these situations generally knew each other as fellow students, as friends or friends of friends, or as professor and student. The experimental situation was not based on the interaction of strangers.

What we find in Wundt's Leipzig laboratory is a distinctly collaborative style of interaction among the participants in the experimental situation, not a situation characterized by fixed role and status differentials. (Of course, Wundt himself occupied a special position, but once he entered the experimental situation as a subject his responses were tabulated along with everyone else's and were not given any special importance.) In these early experiments the interaction of investigators with their human data sources has not been clearly separated from the interaction among the investigators. Most of the individuals who functioned as subjects in these experiments were members of the research community. Not all the subjects who appear in the research reports of the Leipzig group were students of psychology, but those who were not seem to have had a definite interest in the subject and generally some experience of psychological experimentation. So instead of a two-tiered social structure in which there is a clear distinction between one social interchange with experimental subjects and another with colleagues, there is a less differentiated situation in which, to a large extent, the scientific community is its own data source.

This style of psychological experimentation appears to have resulted from the convergence of a particular tradition of scholarly work with a special kind of research goal. The scholarly tradition was that associated with the German nineteenth-century university system. It was based on linking the training of an intellectual elite with the systematic production of new knowledge in the context of a collaborative research enterprise.[6] The research goal involved the analysis of generalized processes characteristic of the normal, mature human mind. Psychological experimentation was designed to investigate this object much as physiological experimentation was designed to analyze biological processes involved in the functioning of normal, mature organisms.

But just as the practice of introspection had helped to construct the object it was meant to investigate, so the new practice of psychological experimentation constructed its own object. Experimental subjects were not studied as individual persons but as examples that displayed certain common human characteristics. That is why the role of subject could be assumed by any member of the research community. They did not represent themselves but their common mental processes. These "elementary" mental processes, as Wundt called them, were assumed to be natural objects that could be studied independently of the whole personality. All that was necessary were the restricted conditions of the laboratory and a certain preparation of the subject. That preparation involved a specific kind of background and a suitably collaborative attitude. But this bringing together of restricted conditions of stimulation and response – of psychologically sophisticated background and of a readiness to play the subject role – amounted to a careful construction of the very object that the research program was designed to investigate. This did not make it any less "real," but it did mean that it would have to compete for attention with other psychological objects constructed in different ways.

Alternative models of psychological investigation

The model of experimentation inaugurated at Leipzig had a special significance for the early development of psychology as an academic discipline with scientific pretensions, but it was not the only model of psychological experimentation available at the time. The *clinical experiment* had goals and social arrangements of quite a different order. The first major research program that depended on this type of experimentation involved the experimental study of hypnotic phenomena. At exactly the same time that Wundt's Leipzig laboratory was getting under way, a group of French investigators embarked on the systematic use of experimental hypnosis as a tool of psychological research.[7] In their studies various psychological functions were investigated under conditions of experimentally induced hypnosis. Now, contrary to the practice in Leipzig, there was no interchange of experimental roles among the participants in these French studies but a clear and permanent distinction between experimenters and individuals experimented on. Experimenters remained experimenters and hypnotic subjects remained hypnotic subjects. Moreover, there was a glaring status difference between these male scientists and their generally female lay subjects. The distribution of social functions approximated the pattern of the typical modern experiment, with the function of providing the data source being strictly segregated in an experimental subject role.[8]

By 1890, Binet had turned from experiments on hypnotized subjects to experiments on infants.[9] This was possible without altering the essential social structure of the experimental situation. For the Wundtian experiment, on the other hand, this kind of extension of scope was not possible.

The clinical experiment had emerged in a medical context. Those who functioned as subjects in these experiments were identified by labels such as "hysterics" or "somnambulists." Where normal or "healthy" subjects were used, it was for purposes of comparison with diagnosed clinical cases who were the essential target of the research. The experimenters were individuals with a medical background. Before experimental sessions began, the experimenter and the subject were already linked in a physician–patient relationship, and the essential features of this relationship were simply continued into the experimental situation. The whole situation was defined in medical terms. A crucial feature of this definition was the understanding that the psychological states and phenomena under study were something that the subject or patient underwent or suffered.

In the context of the clinical experiment, we find the first consistent usage of the term "subject" in experimental psychology. These medically oriented experimenters quite spontaneously refer to the patients they experimented on as "subjects" (*sujet*), because that term had long been in use to designate a living being that was the object of medical care or naturalistic observation. This usage goes back at least to Buffon in the eighteenth century. Before that a "subject" was a corpse used for purposes of anatomical dissection, and by the early nineteenth century one spoke of patients as being good or bad subjects for surgery.[10] When hypnosis came to be seen as an essentially medical matter, which was certainly the case in the Paris of the 1880s, there was nothing more natural than to extend an already established linguistic usage to yet another object of medical scrutiny. However, within the medical context we immediately get the formulation "healthy subject" (*sujet sains*)[11] when it is a matter of comparing the performances of normal and abnormal individuals. From this it is a very short step to the generalized use of the term subject to refer to any individual under psychological investigation, a step quickly taken by Binet and others.

The English term subject had acquired similar medical connotations as its French equivalent. In the eighteenth century it was used to refer to a corpse employed for anatomical dissection and by the middle of the nineteenth century it could also mean "a person who presents himself for or undergoes medical or surgical treatment" – hence its use in the context of hypnosis.[12] In the English-language psychological literature, the first uses of the term subject occur in the context of experiments involving the hypnotic state.[13] It is interesting to note that at this time Stanley Hall uses the term subject in the context of his work in the area of experimental hypnosis but switches to the terms "percipient" and "observer" when reporting on his experimental work with normal individuals.[14]

J. McKeen Cattell appears to have been the first to use the English term subject in describing a psychological experiment involving a normal adult human data source. However, he is by no means sure of his ground, for in an 1889 paper coauthored by him we find the formulation "an observer or subject," with "subject" in inverted commas.[15] Putting the term subject in inverted commas, when used in this context, was quite common in the English literature of this time, indicating that this usage was of recent origin, possibly influenced by French models.[16]

Thus the earliest years of experimental psychology were marked by the emergence of two very different models of the psychological experiment as a social situation. The Leipzig model involved a high degree of fluidity in the allocation of social functions in the experimental situation. While participants in this situation could assume both the role of experimenter and that of subject, the latter role was, if anything, more important than the former. In the clinical experiment, by contrast, experimenter and subject roles were rigidly segregated. The experimenter was clearly in charge, and only he was fully informed about the significance of the experiment. Experimenter and subject roles were not exchangeable. The objects of study that these two experimental situations were designed to display were clearly different. The clinical experiment was meant to display the effects of an abnormal condition for the benefit of the informed experimenter and those who were able to identify with him. On the other hand, the Leipzig experiment was meant to display universal processes that characterized all normal minds and whose significance was therefore open to all. In the one case the object of investigation presupposed the asymmetry of the experimenter–subject relation, but in the other case it did not.

Although these two models of psychological investigation constituted the types whose fundamental divergence holds the greatest theoretical interest, several other models had begun to emerge before the end of the nineteenth century. The earliest of these first appeared in England. In 1884, at the International Health Exhibition in London, Francis Galton set up a laboratory for "testing" the "mental faculties" of members of the public. Here we have an investigative situation whose social structure is clearly different from that of either the Leipzig or the Paris model. The individuals investigated were not medically stigmatized but presumed to be ordinary members of the public. Yet their role and that of the investigator were definitely not exchangeable, and between these roles there was a clear status difference. The investigator was supposed to have some kind of expert knowledge about the individual tested, and he was willing to share this knowledge upon payment of a fee. Galton charged every person who availed himself of the services of his laboratory the sum of threepence in return for which that individual received a card containing the results of the measurements that had been made on him or her. The relationship operated on a "fee for service" principle, but the service was clearly not thought of as being medical in character. Galton referred to the individuals who presented themselves for testing not as subjects but as "applicants."

There was apparently no lack of "applicants" for Galton's services, over 9,000 having been tested by the time the exhibition closed.[17] This seems to indicate that there was something not altogether unfamiliar about the kind of interaction that Galton's laboratory invited, and that the situation was sufficiently meaningful to a large number of people, not only to induce them to participate but also to pay money for the privilege. Competitive school examinations would certainly have provided a general social model for anthropometric testing. But beyond this it is possible that a more specific social model for Galton's anthropometric laboratory was provided by the practice of phrenology, which, though by then discredited, had been widely relied upon a generation earlier. Phrenologists had long offered individuals the service of informing them about their "mental faculties" on the basis of measurements performed upon them. Some of Galton's "applicants" may not have regarded his service as being so very different from that provided by a familiar model. Even Queen Victoria had consulted a phrenologist about her children, and as a young man Galton himself had gone to a phrenologist and taken his report very seriously.[18] I am not of course suggesting an explicit resolve to imitate the phrenologist's practice, but rather the operation of an analogous social practice as an implicit model that made Galton's innovative procedure immediately acceptable.

A comparison of the social structure of Galton's investigative practice with that of the Leipzig or the clinical experiment reveals some divergent features. The utilitarian and contractual elements of Galton's practice are particularly striking. Galton was offering to provide a service against payment, a service that was apparently considered of possible value to his subjects. What he contracted to provide his subjects with was information about their relative performance on specific tasks believed to reflect important abilities.[19] His subjects would have been interested in this information for the same reasons that they or their parents were interested in the information provided by phrenologists: In a society in which the social career of individuals depended on their marketable skills any "scientific" (i.e., believed to be objective and reliable) information pertaining to these skills was not only of possible instrumental value to the possessors of those skills but was also likely to be relevant to their self-image and their desire for self-improvement.[20]

From Galton's own point of view the major hoped-for return on this large-scale investment in "anthropometric measurement" was certainly not the money but a body of information that might ultimately be useful in connection with his eugenics program. The practical implementation of this social program of selective human breeding would have been facilitated by a good data base on human ability. Galton's interests in this research situation were just as practical as those of his subjects, the difference being that while they were interested in their plans for individual advancement he was interested in social planning and its rational foundation.

Three models of practice compared

The object of investigation that Galton and his subjects created through their structured interaction in the testing situation was very different from the objects constituted by either the Leipzig or the clinical experiment, although it had elements in common with both. What Galton's testing situation produced was essentially a set of individual *performances* that could be compared with each other. They had to be *individual* performances – collaborative performances were not countenanced in this situation. At least the performances were defined as individual, and the fact that they were the product of a collaboration between the anthropometrist and his subjects was not allowed to enter into the definition. The performances therefore defined characteristics of independent, socially isolated individuals and these characteristics were designated as "abilities." An ability was what a person could do on his own, and the object of interest was either the individual defined as an assembly of such abilities or the distribution of performance abilities in a population. The latter was Galton's primary object of interest, the former that of his subjects.

Locating the object of research within an isolated individual person was of course a basic common feature of all forms of psychological investigation; in fact, this constituted it as psychological, rather than some other kind of investigation. However, the line of inquiry inaugurated by Galton was the most uncompromising in this respect. As we saw in the previous chapter, Wundt was uncomfortable about the way in which the experimental method isolates the individual from the social–historical context, so that he limited the reach of this method to the "simpler" mental processes. The French clinical experimentalists were interested in the interindividual process of suggestion, which they mostly regarded as a fundamental condition for the appearance of the abnormal states they wanted to investigate. But Galton's anthropometry was quite radical in its conceptual severing of the links between an individual's performance and the social conditions of that performance. It accomplished this by defining individual performances as an expression of innate *biological* factors, thereby sealing them off from any possibility of social influence.

The potential practical appeal of Galton's anthropometric method was based on three features. First was the radical individualism, which we have just noted meant that it claimed to be dealing with stable and unalterable individual characteristics that owed nothing to social conditions. This provided the necessary ground for supposing that the performance of individuals in the test situation could be used as a guide to their performance in natural situations outside the laboratory.

Second, there was the partly implicit element of interindividual competition built into the method. Of interest was not just one individual's performance but the relative standing of that individual in comparison with others. For this type of investigative practice any one episode of experimenter–subject interaction only obtained its significance from the fact that it was part of a series of such interactions, each of them with a different subject. In the other forms of investigative practice single laboratory episodes were also often part of a series, but here the series was constituted by a variation of conditions with the same subject. In the Galtonian approach the series of experimental episodes was essentially defined by the differences among the subjects. Of course, duplication of an experiment with different subjects was known both in the Leipzig laboratory and in clinical experimentation. But this was merely a replication of the experimental episode for the purpose of checking on the reliability of the observations made. For the practice of anthropometry, on the other hand, the set of experimental episodes with different subjects formed a statistically linked series, and it was this series, rather than any of the individual episodes, that formed the essential unit in this form of investigative practice. Only in this way was it possible to generate the desired knowledge object: a set of performance norms against which individuals could be compared.

Whereas the Leipzig style of psychological experimentation only made use of the division of labor between experimental and subject roles, the Galtonian approach added to this the multiplication of subjects as an inherent and necessary component of its method. This was closely connected to a third feature of its practical appeal, namely, the statistical nature of the information it yielded. To be of practical value, the comparison of individual performances had to be unambiguous, so that rational individual or social policy decisions could be based on it. The way to achieve this was by assigning quantitative values to performances, thus allowing each individual performance to be precisely placed in relation to every other performance. Therefore, the anthropometric approach was designed to yield knowledge that was inherently statistical in nature. For the Galtonian investigator the individual subject was ultimately "a statistic." This made him a very different kind of knowledge object than the "case" of clinical experimentation or the exemplar of a generalized human mind, as in the Leipzig experiments.

These differences in the status of the knowledge object were linked to certain features of the interaction between investigators and their human subjects. What was of particular relevance here was the bearing of the prior relationship between experimenters and subjects on the social system of experimenter–subject interaction. In Galtonian practice investigators and their subjects met as strangers, contracting to collaborate for a brief period of experimentation or testing. In the clinical experiment investigators and subjects generally extended an existing pattern of physician–patient relationships into the experimental situation. In the Leipzig laboratory, as we have seen, the participants were fellow investigators whose common interests and friendships often extended across several experiments.

Thus, in each case there was a certain fit between the kind of knowledge object sought for and the nature of the experimental interaction that was used to produce it: The superficial interaction of strangers in a limited, contractually defined relationship was the perfect vehicle for the production of a knowledge object whose construction required a series of brief encounters with a succession of human subjects.[21] Similarly, the collaborative interaction of colleagues in exchangeable subject and experimenter roles was nicely suited to the construction of a knowledge object defined in terms of common basic processes in a generalized adult mind. Finally, the phenomena produced by clinical experimentation depended in large measure on a tacit agreement among the participants to extend the medical context into the experimental situation.

Each type of investigative situation constitutes a coherent pattern of theory and practice in which it is possible to distinguish three kinds of interdependent factors. First of all, there is the factor of *custom* which provides a set of shared taken-for-granted meanings and expectations without which the interaction of the participants in the research situation would not proceed along certain predictable paths. Much of this pre-understanding among the participants depends of course on general cultural meanings, but what is of more interest in the present context are the local variations in these patterns. We have distinguished several varieties of custom that played a role in the emergence of independent styles of psychological investigation in different places. These were the customs of the German nineteenth-century university research institute, of scientific medical investigation, of competitive school examinations, and possibly of phrenological assessment. Here were the sources of models of social interaction that could be readily adapted to the purposes of psychological research. The existence of these models meant that psychological research could get started without having to face the impossible task of inventing totally new social forms. However, once it did get successfully started the new practice led to new variants of the old forms and established its own traditions.

The actual practices of psychological investigation, though derived from certain customary patterns, were not identical with these patterns and therefore constituted a second distinguishable element in the emerging local styles. A third element was constituted by the diverging knowledge interests that have already been alluded to. The three original forms of psychological research practice were by no means interested in the same kind of knowledge. There was a world of difference between knowledge of elementary processes in the generalized human mind, knowledge of pathological states, and knowledge of individual performance comparisons. From the beginning, psychological investigators pursued several distinct knowledge goals and worked within investigative situations that were appropriate to those goals. However, it would be quite inaccurate to think of their adoption of appropriate investigative situations as a product of deliberate rational choice. The knowledge goals that these investigators pursued were as much rooted in certain traditions as were the practices that they adopted. Intellectual interests and the practices that realized those interests were both generally absorbed as a single cultural complex of theory and practice. Each such complex then led to the construction of the kind of knowledge object it had posited.

Modern psychology began with several different models of what a psychological investigation might look like, and the differences among these models went rather deep. Psychological investigation only existed in a number of different historical incarnations. Although in certain locations a particular model of investigation might achieve an overwhelming predominance for a considerable period, this should not mislead us into equating a particular form of experimentation with experimentation as such. Questions about the scope and limits of experimentation in psychology have usually presupposed a particular variant of experimental practice. However, it is historically more meaningful to compare the implications of different forms of practice.

Social psychological consequences

One set of questions that is suggested by the historical analysis of psychological experiments concerns the social psychology of such experiments. If there were different models of psychological investigation, with different patterns of social interaction among the participants, then each such model is likely to have entailed its own social psychological problems. In historical perspective it is inadequate to take one type of investigative situation as

Figure 4.1. Wilhelm Wundt and assistants at the Leipzig laboratory (photograph ca. 1910)

Figure 4.2. A medical demonstration of hypnosis and grand hysteria; from a painting by André Pierre Brouillet, *The School of Jean-Martin Charcot*

Figure 4.3. Galton's anthropometric laboratory in London

representing *the* psychological experiment and to generalize from the social psychological factors in this situation to all types of psychological experimentation.

One also needs to distinguish between the intended and the unintended social features of experimentation. This is where the term "artifact" can be seriously misleading.[22] All results of psychological experiments are social artifacts in the sense that they are produced by the participants in certain specially constructed social situations. But insofar as this production corresponds to the intentions and plans of those in charge of the experiments, it is not thought of as artifactual. That epithet is commonly reserved for the unavoidable *unintended* social features of experimentation – for example, the suggestive influence of the experimenter's hopes and expectations, or the inappropriate anxieties of the subject. However, just as the planned social features of experimentation vary considerably in different models of the experimental situation, so do the unplanned features. It is not unlikely that each basic model would have been prey to a different set of unintended social psychological disturbances.

In the clinical experiment a major source of such disturbances would have to be sought in the profound status differential that rigidly separated experimenters and subjects. This status differential meant that the experimental situation was, among other things, an encounter between individuals who were very unequal in power. Moreover, this power inequality was an *essential* feature of the experimental situation, because the subject's role was that of providing the material *on* which the experimenter could make his observations and test his hypotheses. The subject had to play the role of a biological organism or medical preparation if the experimental scenario was to be completed successfully. Finding oneself in this inferior position, the subject could be expected to be affected by the stresses and anxieties that people in such positions commonly experience. These subjects might have been expected to experience some anxiety about receiving a favorable evaluation from the experimenter and to worry about possible harmful consequences to themselves. A situation that they did not control and only half understood would have aroused anxiety in any case.

However, from the viewpoint of the outcome of the experiment, these anxieties were probably less serious in themselves than the various coping strategies in which the subjects, as persons in inferior and relatively powerless positions, could be expected to indulge.[23] These strategies might vary all the way from apathetic reactions to strenuous collusion with the investigator to give him the kind of performance he wanted. In any particular case one would not know how much of what was displayed to the investigator was due to peculiarities of specific coping styles and how much was due to the factors in which the investigator was really interested.

At this point we should not overlook the fact that the social structure of experimental situations may have had unintended social psychological consequences for the experimenter's as well as for the subject's role. People in positions of power and control are quite apt to drift into arrogance and callousness in the absence of appropriate institutional safeguards. Insofar as the social situation of the experiment is marked by a clear power differential, the psychological reactions of the more and the less powerful role incumbents are likely to be complementary to one another. So the performance the subject puts on for the benefit of the investigator is not recognized for what it is because the investigator does not think of the subject as a human agent but simply as material for the demonstration of his pet ideas. The extreme illustration of this outcome was provided by Charcot's notorious public demonstrations of hysteria at the Salpêtriere.[24] These hardly qualified as experiments, yet the social dynamics that they involved only represented gross forms of processes that had their subtler counterparts in every clinical experiment.

Although the investigative style pioneered by Galton would have shared some of the social psychological problems of the clinical experiment, the less-intense involvement of the participants and the introduction of a certain contractual reciprocity would tend to mitigate these problems. However, the kind of experimenter–subject relationship that this type of study promoted had potential problems of its own. In particular, the divergent interests that experimenters and subjects had in the outcome of the experimental situation, combined with the superficiality of their contact, contained the seeds of trouble. The subject's performance would be rendered in the service of aims that were not those of the investigator and the nature of their relationship was not such as to allow much opportunity for mutual explanations. This was surely a situation in which misunderstanding was likely to flourish. Such misunderstanding would take the form of divergent definitions of the experimental situation, and hence of the experimental task, by experimenter and subject. Although less significant for many of the tasks that Galton employed, it would become a more and more serious problem as Galton's successors increased the psychological complexity and potential ambiguity of experimental tasks.

Participants in the Leipzig model of psychological experimentation were much better protected against problems of this kind, for their collegial relationship and the exchangeability of experimenter and subject roles would ensure that misunderstanding was kept at a minimum. Nor would this style of research be plagued by the psychological consequences of extreme status inequality in the manner of the clinical experiment. However, the Leipzig model was not free of unwanted social psychological side effects of its own. The most noteworthy of these derived from the same source as its strength, namely, the close mutual understanding that developed among the participants. This made it relatively easy for observations to be reported that depended entirely on the conventions and tacit agreements among the members of specific research groups. Such observations were then unverifiable by anyone else. The reports from Titchener's Cornell laboratory on the nonexistence of imageless thought provided the classic example of this outcome.[25] It has been observed that the problems of "introspective psychology" did not arise on the level of comparing the introspective reports of individuals with each other but on the level of reconciling the conflicting claims of different laboratories. Historically, these problems had less to do with the intrinsic limitations of introspection than they had to do with the social psychological context in which introspection was employed – namely, the context provided by the Leipzig model of experimentation.

With the advantage of hindsight, it is easy to see that there was no prescription for arranging the social conditions of psychological experimentation that was free of unintended social psychological side effects. Psychological research was only able to collect its material by putting whole persons into social situations. Even though this research was originally interested neither in the persons nor in the social situation but in certain isolated intraindividual, processes, it could isolate these processes only conceptually. In practice, the research had to contend with the persons who were the carriers of these processes and with the social situations that were necessary to evoke the phenomena of interest. But that inevitably led to the appearance of various side effects, which were unintended and for a very long time not even recognized. However, the precise nature of these effects would vary with the model of investigative practice that was adopted. There was no single unflawed procedure for exposing the truth, only a variety of practices with different sources of error. Any estimate of the relative seriousness of these sources of error would have to depend on the kind of knowledge goal one had adopted.

The eclipse of the Leipzig model

The last three chapters have presented an analytic account of the origins of modern psychology as a form of investigative practice. In order to prepare the transition to the rest of this book, which will deal mainly with twentieth-century developments, let us take a brief look at certain historical developments beyond the earliest period of experimental psychology.

A major development was the rapid adoption of all the major techniques of psychological investigation by American investigators. The Leipzig model of investigative practice was replicated by many of those who had spent some time working with Wundt, most notably by Titchener at Cornell. But what made Titchener somewhat unusual was the fact that he never ceased to insist that the Leipzig model was the only one on which a genuine science of mental life could and should be based. Most of the early American investigators were much less dogmatic about their choice of methods, and some were extremely eclectic. G. S. Hall, for example, carried out different pioneering studies that followed all three of the models we have distinguished.[26] Cattell was particularly active in promoting the style of investigation that had been begun by Galton.[27] Clinical experimentation had its adherents too.[28] Early American psychology presented a mixture of investigative styles with few centers showing exclusive commitment to one style, although some had obvious preferences. However, in the course of the first half of the twentieth century a particular synthetic type of investigative practice developed and came to dominate American psychology. That development will be traced in the following chapters.

The model of investigative practice that proved least viable in the twentieth century was the Leipzig model. Because this model has a specific feature, which is always unambiguously reported in published research reports, it is possible to follow its decline rather accurately by means of a content analysis of psychological journals. The feature in question involves the exchange of experimenter and subject roles among at least some of the participants in the same experiment. Highly characteristic of the Leipzig model, the feature is clearly absent in clinical experimentation and in the type of investigation pioneered by Galton.

Concentrating on this feature as an index of the relative prevalence of the Leipzig model has the practical advantage that it is readily and reliably

Table 4.1. *Percentage of empirical studies in which there was an exchange of experimenter and subject roles*

Journal title	1894–1896	1909–1912	1924–1926	1934–1936
Philosophische Studien	32	79[a]	—[b]	—
American Journal of Psychology	38	28	20	8
Psychological Review	33	28	N.C.[c]	
Archiv für die gesamte Psychologie	—	43	28	16
Zeitschrift für angewandte Psychologie	—	4	N.C.	
Journal of Educational Psychology	—	2	0	0
Psychological Monographs	—	18	18	7
Journal of Experimental Psychology	—	—	4	6
Journal of Applied Psychology	—	—	0	0
Psychologische Forschung	—	—	38	31

[a]The second entry for *Philosophische Studien* is based on the new series of that journal, called *Psychologische Studien*.
[b]Dashes indicate periods before or after the appearance of a journal.
[c]An entry of N.C. means that coding of the journal was discontinued because no significant information was likely to result; for example, in the case of *Psychological Review* the proportion of codable empirical articles went from low to zero.

coded in published research reports. It should be noted, however, that this does not provide an accurate estimate of the *absolute* prevalence of the model, because it misses the cases where research colleagues exchange roles in different experiments but not in the same experiment. This practice was especially prevalent in the early days of experimental psychology, but picking it up from the research literature is time-consuming and hardly justified by the intrinsic interest of the results. There are also other features of the Leipzig model that are missed. However, it is the fluctuation in the *relative* popularity of the Leipzig model over the years that is historically more interesting, and this can be assessed by the simplified index as already described.

Empirical articles published in major American and German psychological journals were therefore coded in terms of the presence or absence of an exchange of experimenter and subject roles in the same study. (The sampling of journal volumes for different time periods, and the rules governing the inclusion of articles in the coding process are described in the Appendix.) Table 4.1 shows the results of this content analysis.

Although the simplified index used underestimates the use of the Leipzig model, this underestimation tends to be greatest when the use of this model is relatively high, especially in the earliest period. So the simplification in coding dilutes rather than accentuates the overall trend. This trend nevertheless emerges quite clearly. It involves an overall decline in the practice of exchanging experimenter and subject roles. Where continuity of publication allows a comparison of the same journal over several decades, the practice generally declines steadily with each time sample. The best documented case is that of the *American Journal of Psychology* where incidence of the practice gradually declines from 31 to 8 percent over a forty-year period. These figures do not suggest a sudden abandonment of the practice so much as a steady deterioration in its relative frequency of adoption. Nor does this general trend appear to be a specifically American phenomenon, because among the German journals, the one that is not associated with a particular school, *Archiv für die gesamte Psychologie,* shows the same trend.

It does seem as though the kind of experimental practice represented in this index was always somewhat more popular in German than in American psychology. In any time period the German figures are always higher than the American. By far the highest incidence (79 percent) of this pattern of research is to be found in *Psychologische Studien,* the Leipzig successor to Wundt's *Philosophische Studien.* But the Gestalt journal, *Psychologische Forschung,* also maintains a relatively high incidence on the index at a time when the practice of exchanging roles is being virtually abandoned by everyone else. At no time after 1890 did the original Leipzig style of arranging psychological experiments really dominate the discipline. At best it was one model among others. Never dominant, it had dwindled virtually to insignificance in American psychology during the period preceding World War II.

For applied psychology, whether American or German, it never had the slightest appeal, as is shown by the figures for the relevant journals, *Zeitschrift für angewandte Psychologie, Journal of Applied Psychology,* and *Journal of Educational Psychology.* Applied psychology had committed itself to knowledge goals that were unlikely to be advanced by the kind of investigative practice associated with Wundt's laboratory. What it was after was knowledge that could be quickly utilized by agencies of social control so as to make their work more efficient and more rationally defensible. Knowledge that led to behavioral prediction suited this purpose, but knowledge obtained in situations where the participants collaboratively explored the structure of their experience did not. Thorndike, a major pioneer of the new applied psychology, saw this very clearly when he complained that some of the procedures of academic experimental psychology "were not to aid the experimenter to know what the subject did, but to aid the subject to know what he experienced."[29] The kind of knowledge that Thorndike and others who shared his interests were after was predicated on a fundamental asymmetry between the subject and the object of knowledge. This asymmetry was not merely epistemic, however, but also social, because it was accepted that it was to be useful to the subject (or those for whom he acted) in controlling the object. What was desired was knowledge of individuals as *the objects of intervention rather than as the subjects of experience.* Experimental situations in which there was an exchange of roles between experimenters and subjects might well have seemed irrelevant to this type of knowledge goal.

As more and more academic psychologists, especially in the United States, adopted knowledge goals that were similar to those that had characterized applied psychology from the beginning, the Leipzig model was pushed even further into the background. It survived, on a relatively much reduced scale, only in the area of research on sensation and perception, the area in which it had also registered its first successes in the previous century.[30] Thus, as psychology's knowledge goals changed, its preferred mixture of investigative practices changed too.

Neither clinical experimentation nor the style of investigation pioneered by Galton was left unaffected by these changes. On the contrary, the Galtonian approach in particular underwent a dynamic development during the first half of the twentieth century. But this affected the nature of the knowledge produced rather than the social parameters of the experimenter–subject relationship. To analyze these developments, we will have to turn from the social interior of the investigative situation to its social exterior – that is, to the relations of investigators with each other and with the wider society. The making of a psychological knowledge product involved more than just the interaction of experimenters and subjects. It involved matters of conceptual form, of appropriate packaging, and of dissemination through approved channels. Twentieth-century developments in investigative practice mostly affected these aspects, and it is to them that we must now turn.

5

The triumph of the aggregate

Knowledge claims and attributions

Modern psychology may have started out with at least three different types of investigative practice, but, as we have seen, one of these hardly survived the first few decades in the history of the discipline. Among the other two, it was the Galtonian type that was to achieve a dominant position during the first half of the twentieth century. In the following chapters we will have to analyze various aspects of this complex historical development. Our analysis will draw heavily on the information contained in the psychological journals that now provide the major vehicle for the establishment of psychological knowledge claims. Twentieth-century psychology is a system of "public knowledge,"[1] and the major disciplinary journals constitute the crucial link in the transformation of the local knowledge produced in research settings into truly public knowledge.

This transformation involves the preparation of a research report that meets the criteria and conventions that prevail within the discipline or research area. Ostensibly, the research article communicates certain "findings" held to represent knowledge by the conventions of a particular community. In fact, the contents of the article only represent a knowledge *claim,* which may or may not be accepted by its intended audience. To convince this audience, the authors must submit to its conventions. For instance, they must adopt a certain literary style, relate the contents of the article to known concerns and interests of its possible audience, demonstrate the conformity of their methods to prevailing norms, and so on.[2] But the centerpiece of its work of persuasion is likely to involve the "findings" themselves, for they are after all the raison d'être of the whole paper. If the "results" that the author(s) wish to communicate do not measure up to prevailing standards of what constitutes significant knowledge in their chosen field, the report will not see the light of day, no matter how immaculate its style might otherwise be. These standards, however, are not eternal verities but are historically variable. Individual researchers must take these standards as they find them, but the potential contribution of a historical perspective lies in uncovering their contingent nature.

Whereas sociological field work has been able to study the way in which individual researchers accommodate themselves to prevailing standards of knowledge production, historical analysis examines changes in these standards themselves. This can be done by following changes in the patterns of reporting practices in selected journals over a period of time.

To establish their knowledge claim, researchers must transform whatever records they have made during the process of investigation into a standard format that is acceptable in their field of work. That is to say, they must "decontextualize" their records of what they did and saw in the laboratory and "recontextualize" this material in accordance with the standards of the audience that will adjudicate their knowledge claim.[3] In the case of psychological research certain aspects of the participants' interaction must be transformed into a symbolic form, which will eventually appear as the "findings" of the investigation. It is this form we now need to examine.

When one examines knowledge claims, as reflected in scientific psychological papers, one finds that they always have a basic propositional form. Some set of attributes is said to be true of certain subjects under certain conditions. What the "results" section of the paper says is that one or more persons did such and such when in a particular situation. The knowledge objects that are the legitimate focus of interest for scientific psychology involve the attributes of a specially constituted subject, namely, the research subject. Most commonly, it is the attributes, rather than the subject whose attributes they are, that are considered to be problematic for psychological research. But if we are to achieve any insight into fundamental changes in the investigative practices of the discipline, we need to reverse this perspective and to recognize the problematic status of the research subjects themselves. For when the nature of the subjects changes, the nature of their attributes changes too. For instance, a subject who is defined as an individual consciousness has a different set of attributes from a subject who is defined as a biological population. If we want to understand the most general historical changes in research practice, we need to deconstruct the research subjects that an army of investigators have constructed in obedience to various conventions about the proper objects of psychological knowledge. Much of the remainder of this book will be devoted to this analysis.

Individual attribution of experimental data

A systematic survey of experimental psychological journals (*American Journal of Psychology, Psychological Review, Philosophische Studien, Psy-*

chologische Studien, Archiv für die gesamte Psychologie) for the period up to World War I shows that in virtually all published studies the experimental results were clearly attributed to individual experimental subjects. Most often the individuals concerned were identified by a name, letter, or initials, but the point worthy of attention is that the form of reporting experimental data involves the attribution of certain "responses" to specific individuals. This pattern is best illustrated by an example from the psychological literature of this period (Table 5.1). Even where results were averaged across a number of subjects, which was often not the case, the responses of each individual were systematically reported. Typically, it was the pattern of individual responses, and not the average result, that formed the basis of the theoretical discussion.

How was this pattern of reporting experimental results related to psychological knowledge claims? Even where the table of results only identified the human data source by a letter or initial, it was usually possible to verify, from other sections of the experimental report, that the individuals to whom the reported responses were attributed were in fact competent and experienced psychological "observers" whose testimony had much greater value than that of naïve subjects. To some extent the knowledge claim undoubtedly rested on the reliability attributed to expert observers. Attaching the name of a reputable observer was a way of establishing the scientific credibility of the data.

However, the individually identified expert research subject could only function in this manner because of the way in which the object of psychological knowledge was defined. In this research tradition it was unquestioned that psychological knowledge had to pertain to the *content of individual minds*. In other words, such knowledge could always be expressed in propositions of which the individual mind was the subject and various statements about mental contents were the predicate. The arrangement of tables of experimental results simply reflected this schema of psychological knowledge by attributing a particular pattern of mental contents to specified individual minds. This represented a new way of constituting psychology's traditional knowledge object, the individual consciousness. Thanks to the new methods, that apparently private object was transformed into an object of public knowledge.

Individual attributions of experimental results did not mean that the phenomena studied were felt to be idiosyncratic. On the contrary, claims for the general or even universal significance of experimental findings were commonly made, but such claims were generally based on one or more additional considerations. For instance, in such areas as the psychology of sensation results from even a single subject could be claimed to have general significance because of the presumed similarity of the underlying physiology in all normal human individuals. Second, claims for generality were supported by repeating the experiment with a few individuals. Each additional

Table 5.1. *A table of results from the* American Journal of Psychology, *1895*

	OBSERVER, DR. KIRSCHMANN.						OBSERVER, J. O. QUANTZ.					
	LEFT EYE.			RIGHT EYE.			LEFT EYE.			RIGHT EYE.		
Division of Exps.	No. of Single Exp.	Av. Deviation from Norm. Mag (o ' ")	M. V. (' ")	No. of Single Exp.	Av. Deviation from Norm. Mag (o ' ")	M. V. (' ")	No. of Single Exp.	Av. Deviation from Norm. Mag (o ' ")	M. V. (' ")	No. of Single Exp.	Av. Deviation from Norm. Mag (o ' ")	M. V. (' ")
Equal Discs	50	+0 2 8.7	1 14.92	50	+0 1 10.2	0 40.73	50	+0 0 46.55	0 34.9	50	+0 0 28.7	0 57.05
Unequal Discs	50	+0 2 53.8	1 41.8	50	−0 0 5.5	1 19.25	50	+0 4 58.8	1 16.65	50	+0 3 41.5	1 22.28
Av. for Single Eyes	100	+0 2 31.25	1 28.36	100	+0 0 32.35	0 59.99	100	+0 2 52.68	0 55.78	100	+0 2 5.1	1 9.66

	No. of Single Trials.	Average Deviation.		M. V.
		Absolute Value. (' ")	Val. Rel. to Norm. Mag.	(' ")
Total Average (Kirschmann)	200	+1 31.8	+0.01355	1 14.18
Total Average (Quantz)	200	+2 28.89	+0.0221	1 2.72

Note: This table is from J. O. Quantz, "The influence of the color of surfaces on our estimation of their magnitude," *American Journal of Psychology* 7 (1895):33. The table is reproduced here merely as an example of personal attribution to named individual observers; the content is not relevant in the present context.

subject then provided material for replication of the original experiment. Third, where attempts at replication were not altogether successful, relevant personal information or introspective evidence could be used to provide a rational explanation of the failures; for example, relevant previous learning experiences or reported fluctuations of attention could be used to account for discrepant results.

There were essentially two versions of this first model of experimentally established psychological knowledge, a strong and a weak version. The strong version was due to the inspiration of Wundt for whom the individual mind was a synthetic unity constituted by real processes of "psychic causality." Mental contents were the attributes of this system and could be used to study its constitution. In the weaker version of some of Wundt's pupils, like Titchener, the reference to real causal processes was dropped, but the individual mind remained the only conceivable locus for mental contents. In fact mental contents retained their status as that which constituted individual minds, although the nature of this constitution was interpreted positivistically rather than in the terms of Wundt's principle of "psychic causality."[4] Throughout this research tradition Wundt's *actuality* principle continued to be respected in principle, if not always in name. That is to say, knowledge of mental contents was interpreted as knowledge of actually occurring events rather than as logical schemes imposed on the products of such events. Whereas the latter approach would not necessarily have required any reference to individual minds, the study of mental contents in their actual occurrence was hardly conceivable without such individual attribution.

Parenthetically, it should be noted that there continued to be a certain analogy between this tradition of psychological experimentation and the tradition of physiological experimentation out of which it had grown. What marked off the relevant domain of physiological experimentation from biochemical or biophysical studies on the one hand, and from biological population studies on the other, was its restriction to the universe of the individual organism.[5] In its most consistent version, this tradition even entailed a commitment to experimentation involving the whole organism. This commitment was explicit in the work of Pavlov, for example, whose experiments on conditioned reflexes provided a kind of objectivistic analogue to the experimental work of mentalistic psychology. For Pavlov the individual living cerebrum was the unit of study, just as the individual active mind was for the experimental psychologists.[6] Both for Wundt and for Pavlov this methodological feature was not just a practical preference. They both saw the purpose of their experiments in terms of an exploration of a very specific system of causal processes, although for Wundt the system was one of psychic causes, whereas for Pavlov it was one of physiological causes. Pavlovian experimentation illustrates the lack of any reciprocal necessity in the connection between the mentalism of early experimental psychology and the individualism of its knowledge claims. Mentalistic knowledge interests may predispose one to individual attribution in one's knowledge claims, but the converse is decidedly not true.

Also, the individual attribution style of making knowledge claims has nothing to do with whether the features that are individually attributed are quantitative or qualitative in nature. In fact, most of the literature with which we are concerned here involves quantitative data. Moreover, individual attribution does not by any means exclude the use of statistics. Systematic psychological experimentation involved use of statistics from the very beginning. This is illustrated by experiments in psychophysics, which were of course a mainstay of the early psychological laboratory. Here, statistical methods were used as a way of combining observations in order to derive an assumed true value still attributable to an individual observer. The calculus of error was applied to a population of observations from a single subject, not to a population of subjects. Any increase in the number of experimental subjects above one constituted a replication of the experiment. If the interindividual variability was large, this was considered prima facie evidence that the attempted isolation of critical determining factors had failed and that uncontrolled disturbing processes had supervened.[7]

Although this early research tradition of modern psychology was constituted by a set of coherent and internally consistent practices, it was apparently inadequate to the changing tasks that psychologists set themselves in the course of the twentieth century. The importance of this style of research practice for the discipline as a whole declined quite rapidly during the early part of the century. At the end of chapter 4 we documented the decline of the Wundtian research paradigm by using the exchange of experimenter and subject roles as a rough index. Before proceeding further we will illustrate this trend by another index that arises out of the preceding analysis of research reports. As we noted in the course of this analysis, the individual attribution of research data was frequently buttressed by identifying experimental subjects by name. We can therefore use the relative prevalence of this practice as an index to demonstrate the progressive decline of this method of establishing psychological knowledge claims. The relevant data are presented in Table 5.2.

It is apparent that the general historical trend parallels the trend illustrated in Table 4.1. In the course of time, experimenter and subject roles are less and less frequently exchanged and research subjects are less and less frequently identified by name. What factors were involved in this decline? That is obviously a very large question, and it cannot be answered by considering the declining research style in isolation from its competitors. The other side of the coin in the relative decline of one set of practices is obviously represented by the relative ascendancy of another set of practices. What we would like to understand is the replacement of one kind

Table 5.2. *Percentage of empirical studies in which all or some of those participating as subjects are identified by name*

Journal title	1894–1896	1909–1912	1924–1926	1934–1936
American Journal of Psychology	54	39	39	24
Psychological Review	35	35	N.C.[a]	
Journal of Educational Psychology	—[b]	4	0	0
Psychological Monographs	—	18	21	5
Journal of Experimental Psychology	—	—	8	8
Journal of Applied Psychology	—	—	4	0

[a]The N.C. entry for the *Psychological Review* indicates the point at which the number of empirical studies published in this journal became too small for meaningful comparison.
[b]Dashes indicate periods before the appearance of a journal.

of investigative practice by another. But this means we have to explore the alternatives more extensively than we have done thus far. One potential basis for an alternative style of psychological research was provided by the pioneering work of Galton, which involved the construction of a different kind of knowledge object whose historical development now needs to be considered more fully.

Alternative basis for knowledge claims: The psychological census

The alternative to the individual attribution of psychological data is to be found in attempts at making psychological knowledge claims by attributing psychological characteristics to collective rather than individual subjects. In other words, one could claim that certain psychologically interesting features were true of some aggregate of individuals. This meant that one's knowledge claims necessarily took a statistical form, the comparison of different aggregates usually being necessary to arouse interest in the data. One could compare different age groups, for example, on some psychological characteristic. Early forms of this sort of statistical comparison were crude, being generally limited to averages, or occasionally, ratios of the population meeting some criterion. The important point was that these research reports would attribute psychological characteristics to groups of subjects rather than to individuals.

Whereas the dominant form of psychological research reporting seemed to take experimental physiology as its model, this alternative style of presenting the results of psychological research appeared to owe more to a practice of quantitative social research that had emerged earlier in the nineteenth century. Indeed, statistical studies of human conduct in the aggregate were both longer and better established than were laboratory studies of psychological processes. Interest in the statistics of crime, suicide, and poverty and in indexes related to public health was quite widespread before the middle of the nineteenth century, both in Europe and in North America. So-called statistical societies, which typically linked the compilation of social statistics to questions of social reform, were active in a number of countries. The numerical depiction of social problems seemed to promise a more "scientific" approach to their solution than the road indicated by political ideology.[8]

On the practical level, the development of social statistics was closely linked to the spreading use of the questionnaire as a method of investigation. Rather than rely exclusively on official statistics the statistical societies began to collect their own information about the living and working conditions of the poor. It was a small step from circulating questionnaires about conditions of child labor to the use of such instruments for the investigation of children in schools.[9] In Berlin, a municipal statistical office was systematically using questionnaires for collecting mass data on children by the 1860s.[10] G. Stanley Hall used the Berlin work as a model in his earliest American investigations of "the content of children's minds."[11] A converging development in the field of medical statistics had led to the compilation of psychological information on children by circulating questionnaires to mothers.[12]

The general popularity of descriptive statistical inquiry in the third quarter of the nineteenth century had made the questionnaire method seem like an appropriate tool for investigating questions of an undoubtedly psychological character. Charles Darwin used it in his study of emotional expression, and his cousin, Francis Galton, used it in his study of heredity and mental imagery.[13] Such usage helped to strengthen the legitimacy of statistical data compiled on the basis of questionnaires as a source of scientific knowledge. This was particularly true of those, like G. Stanley Hall in America, who saw psychology in quasi-Darwinian terms. From this point of view it was important not to restrict psychology to the study of individual minds but to extend it to the distribution of psychological characteristics in populations. Hall gave practical expression to this attitude in the area of child study, of which he was the chief American promoter. In this field the attribution of psychological characteristics to populations, often quite large populations, was common.[14]

But what made the questionnaire acceptable as a tool of psychological research, at least in some quarters, was not just its practical convenience. There was a belief that mass data, gathered by these means, constituted a valid basis for psychological knowledge. This belief was quite new in the latter part of the nineteenth century, and it had needed some profound social and conceptual developments to make it appear plausible. Let us briefly trace some salient aspects of the conceptual background.

The scientific appeal of statistical tables depended on more than the reduction of complex issues to countable "facts." It depended on the numerical regularities that began to appear when the actions of many individuals were aggregated. Rates of crime and suicide, for instance, tended to remain relatively constant for a particular area, and variations among areas could be related to identifiable environmental influences. It was not experimental psychology but the repeated demonstration of striking regularities in social statistics that first convinced a large public that human conduct was subject to quantitative scientific laws.[15] The major methodological implication of these highly effective demonstrations was that the inherent lawfulness of human conduct would become apparent only if observations on a large number of individual cases were combined. This led to an infatuation with large samples, for only through them it seemed could the laws governing human action be made manifest. A felicitous convergence of theoretical and practical interests now became apparent. The collection of large-scale statistical information on human populations had originally been prompted by the administrative concerns of public bureaucracies. Those concerns had dictated an interest in large numbers, and now it began to appear as though large numbers were also necessary to establish the scientific laws of human conduct. Social statistics united the interests of science and administration in a mutually convenient bond, a union that, as we shall see in chapter 7, was to be perpetuated in the psychological statistics of the early twentieth century.

The road that led from social to psychological statistics depended on two fundamental conceptual steps. Quetelet,[16] the pioneer of a statistical social science, took the first step. In order to explain differential but stable rates of social indexes like crime, he invented so-called penchants or propensities which were attributed to an average individual. Thus, differential rates of crime depended on variations in the average "propensity to crime." If rates of crime varied with such factors as age, sex, and climate, this was due to the influence of these factors on the propensity to crime. By analogy, one could think of propensities to suicide, homicide, and insanity. For every such legally or administratively defined category one could imagine a corresponding "propensity." What Quetelet had done was to substitute a continuous magnitude for the distinct acts of separate individuals. Suicide, crime, homicide, and other social acts were not to be understood in terms of the local circumstances of their individual agents but in terms of a statistical magnitude obtained by counting the number of heads and the number of relevant acts in a particular population and dividing the one by the other. The ultimate significance of this step for the methodology of modern psychology cannot be overemphasized.

However, it took a considerable time for this kind of statistical thinking to be directly applied to psychological questions. At first, there was considerable divergence of opinion about what the regularity of social statistics implied for individual conduct. Quetelet attributed the "propensities" to an average individual, not to specific individuals,[17] and German nineteenth-century statisticians argued strenuously against making inferences of individual determination on the basis of regularities in social statistics.[18] In England, on the other hand, a different view rapidly gained ground. Buckle, who was very widely read, was quite insistent that statistical regularities in human populations were clear evidence for the lawfulness of individual actions.[19] Such a claim depended on a very specific view of the relationship between collectivities and their individual members: Not only were individual attributes freely composable into aggregates (implying the continuity assumption), but, conversely, group attributes were to be regarded as nothing but summations of individual attributes. This latter view was consonant with Charles Darwin's contemporaneous conception of a biological species as an interbreeding population of individuals rather than as an essential type.[20]

That such assumptions could provide the basis for a technology of research on individual differences was clearly grasped by Francis Galton. His technical contributions and those of his devoted follower, Karl Pearson, finally made possible a psychology that could plausibly claim to be scientific while not being experimental. A new method for justifying psychological knowledge claims had become feasible. To make interesting and useful statements about individuals it was not necessary to subject them to intensive experimental or clinical exploration. It was only necessary to compare their performance with that of others, to assign them a place in some aggregate of individual performances.[21] Individuals were now characterized not by anything actually observed to be going on in their minds or organisms but by their deviation from the statistical norm established for the population with which they had been aggregated.

As a result of this Galtonian turn, psychology was left with two very different frameworks for justifying its knowledge claims, the one based on an individual attribution of psychological data, the other necessarily relying on statistical norms. Both types of knowledge claim rested on the demonstration of certain regularities that could be interpreted as having psychological significance. But the nature of the regularities was fundamentally different in the two cases. In the case of the traditional paradigm for experimental psychology, as established by Wundt, the regularities demonstrated had an immediate causal significance. One showed how the reactions of an individual psychophysical system changed as the conditions to which it was exposed were varied. The knowledge one obtained referred directly to particular psychophysical systems. But where the established regularities were attributed to groups of individuals, the system to which one's data referred was no particular psychophysical system but a statistical system constituted by the attributes of a constructed collective subject.[22] Whereas the first type of psychological knowledge was similar in principle to the kind of knowledge then obtainable in experimental physiology or even physics, the second type of knowledge seemed to have more in common with census taking and social or biological population studies.

The discrepancy between these alternative kinds of psychological knowledge was enormous. Experimental knowledge claimed its special status on the basis of its similarity to the kind of knowledge on which the older established laboratory sciences had built their success. Its appeal was to its use of skilled observers, unambiguous reactions, and systematically controlled conditions. None of these appeals was available to those who wanted to publish population studies. Quite different criteria for assessing the worth of knowledge claims seemed to operate here. The simple device of multiplying the number of subjects seems to have been resorted to as a compensation for their lack of skill and experience. An intuitive notion of individual deviations canceling out in a large sample probably operated from the start, and in due course this notion was made more precise by the application of the calculus of probabilities.

In the earliest version of the statistical approach to psychological knowledge, the spirit of Quetelet had not been laid to rest. The possibility of deriving *average* values from group data fascinated the early researchers, for these averages were regarded as indexes of some human type, usually an age type or a sex type.[23] The distribution of characteristics in a population was of interest for establishing the existence of common patterns that could be used to define psychological types and subtypes, and also for determining a natural range of observations. Implicitly, the relationship between the individual and the aggregate was still mediated by a type concept, which usually appears in the guise of some supposedly natural, often quasi-biological, category. In this form the statistical approach could hardly compete with the scientific credentials of traditional experimental psychology.

Most psychologists wanted to emulate the harder sciences. Proper scientific knowledge was knowledge pertaining to regular changes in specific psychophysical systems under systematically varied conditions. "Merely" statistical knowledge pertaining to populations was regarded as superficial and incapable of leading to an explanatory science of determinate processes taking place in actual psychophysical organisms.[24] From this point of view the information yielded by population studies was useless for advancing claims to psychological knowledge. The subjects of these studies were children, the most unreliable of observers; their reactions were complex, variable, and unanalyzed, and the conditions under which the information was collected were uncontrolled and open to the influence of unknown factors. Indeed, this kind of study never achieved respectability within the discipline as a whole. Nevertheless, it had some influential supporters within American psychology, including some, like Hall himself and J. McKeen Cattell, who had worked in Wundt's laboratory and were well acquainted with the standard constraints on making psychological knowledge claims. If they gave their support to this alternative style of psychological investigation it was because they were interested in an alternative kind of psychological knowledge. They did not wish to limit such knowledge to an analysis of psychological processes in individual minds but wanted to extend it to include the distribution of psychological characteristics in populations.

For the most part, the early statistical studies established the value of their knowledge claims by an explicit or implicit appeal to a criterion that was unacceptable to traditional experimentalists. This criterion was one of social relevance, of closeness to real life. It operated both for the choice of populations and for the choice of the psychological characteristics attributed to these populations. The groups investigated were defined in terms of culturally or administratively relevant categories, such as boys and girls of a certain age, pupils at certain schools, or Columbia undergraduates. The psychological categories whose distribution was described were also at first taken from everyday life, like the activities of dreaming, playing, reading, writing, and recollecting.[25] Clearly these early population studies sought to legitimate their knowledge claims by appealing to an immediate and obvious practical relevance. They were like a psychological census which, like other kinds of census, might be of use to persons in positions of authority who had to make decisions affecting their charges. Where the appeal of traditional experimentalism was to esoteric scientific norms, the appeal of these statistical studies was to the appearance of practical utility. Except where it appealed through the magical aura of science, the first approach depended on the existence of a rather sophisticated intellectual community, whereas the second approach was more likely to make a direct impact on a lay public.

The grounds for this approach had been prepared by certain historical developments in the second half of the nineteenth century. On the institutional level the educational system had been profoundly transformed into a rationalized system for grading individuals in terms of their ability to measure up to highly standardized social-intellectual demands. This involved new social practices, like standard curricula, age grading, and written examinations, which actually created the kinds of statistical populations that Galtonian psychology took as its basis.[26] The ordering of individuals in terms of their standing in a statistical aggregate was a social task that schools had assumed some time before psychologists were able to show them how to do it better. Of course, for this to be seen as an appropriate task for educational institutions in the first place required specific historical circumstances.

On the ideological level the ground had been prepared by the increasing tendency to conceptualize social problems in terms of populations of individuals. By the closing years of the nineteenth century it was common, especially in the United States, to formulate the human problems of urbanization, industrial concentration, and immigration in terms of the problems of individuals conceived as members of statistical aggregates. Crime, delinquency, feeblemindedness, and so on were easily attributed to the statistical distribution of certain individual characteristics. That meant the transformation of structural social problems into the problems of individuals, which were to be dealt with not by social change but by administrative means. For those psychologists who were prepared to redefine the nature of their subject matter this social context opened up enormous possibilities for establishing a new science that was socially, in the sense of administratively, relevant.[27] For by studying the distribution of psychological characteristics in populations, psychologists might make an important contribution to the management of any social problems that could be plausibly defined in terms of the statistical variation of individual traits.

The young discipline of scientific psychology therefore had to contend with divergent methodological pulls from the very beginning. The norms of laboratory experimentation, as established particularly by Wundt's example, supported psychology's claim to be able to produce knowledge that was comparable with the knowledge produced by the prestigious older experimental sciences. But for this kind of psychological knowledge there was at best a limited demand in the United States.[28] With no apparent practical application, it was difficult to mobilize resources for its support. The kind of knowledge that was in demand had to have clear social relevance, which meant that it had to be applicable to salient social categories and to practical concerns. This kind of knowledge was available, but it could never hope to pass muster before a scientific tribunal. Faced with these opposing demands from an expert and a lay public, the discipline was able, up to a point, to have it both ways. It was sometimes possible to use the appeal of socially relevant psychological knowledge to obtain support for irrelevant laboratory research, and conversely, it was often the case that the magical appeal of laboratory rituals could be used to legitimize false claims to scientific authority by the purveyors of various practical nostrums. But these were essentially short-term expedients that did nothing to improve an inherently unstable situation. In the long term the fortunes of the discipline would depend on its ability to bridge the gap between the two kinds of knowledge claim it was interested in making.

Transition to aggregate data

By 1914, which conveniently marks the end of a period, American psychology contained some rather divergent investigative practices. In particular, there was the opposition between the more traditional attribution of research data to individuals and the more recent practice of attributing it to groups. The question now arises of how these practices were distributed over sections of the discipline and how they fared during the next historical period.

To document these issues, a content analysis of four psychological journals was undertaken covering the period from 1914 to 1936. Articles reporting empirical research on human subjects were categorized in terms of whether the reported data were attributed essentially to individuals or essentially to collective subjects, or whether a mixture of both types of attribution was used.[29]

The five journals from which research reports were sampled included the two most prominent journals of basic experimental research (*American Journal of Psychology* and *Journal of Experimental Psychology*) and the two most important journals devoted to so-called applied research (*Journal of Applied Psychology* and *Journal of Educational Psychology*). As explained in the appendix, all empirical studies that appeared in three consecutive volumes of the journal for each of three time periods were analyzed. These time periods were the mid-1910s, the mid-1920s, and the mid-1930s.[30] Thus, for each of the four journals nine volumes were analyzed in sets of three, spaced roughly a decade apart. The results for each set of three volumes were combined so as to represent a time period rather than a specific year. The percentage of studies using individual, group, or mixed data was then calculated for each time period. The resulting information is presented in Tables 5.3 and 5.4.

Two patterns stand out clearly. The first is a very consistent tendency for group data to appear relatively more frequently in the journals of "applied" research; the second is an overall trend for the use of group data to increase and that of individual data to decrease over the period under review. This latter trend emerges in all five journals, although because the use of group data in the "applied" journals is already very high to begin with, the change is relatively greater in the journals of basic experimental research.

There also appear to be some consistent differences within the group of "applied" journals and within the group of basic journals. The *Journal of Educational Psychology* manifests the overall trend in a slightly more extreme way than the *Journal of Applied Psychology*. In a parallel, though more pronounced way, the *Journal of Experimental Psychology* is always somewhat ahead of the *American Journal of Psychology* in following the prevailing trend. If we use the overall trend as a standard, the *Journal of Educational Psychology* could be characterized as the most progressive of these journals and the *American Journal of Psychology* as the most conservative.

During the historical period between World War I and World War II,

Table 5.3. *Percentage of empirical studies reporting individual or group data: Three journals of basic research*

Type of data	American Journal of Psychology	Psychological Monographs	Journal of Experimental Psychology[a]
1914–1916			
Individual data only	70	62	43
Individual and group data	5	11	19
Group data only	25	27	38
1924–1926			
Individual data only	60	31	41
Individual and group data	6	24	15
Group data only	35	45	44
1934–1936			
Individual data only	31	31	28
Individual and group data	14	16	11
Group data only	55	53	61
1949–1951[b]			
Individual data only	17	3	15
Individual and group data	3	3	2
Group data only	80	94	83

[a]The *Journal of Experimental Psychology* only began to appear in 1916; the analysis for 1914–1916 is based on the first three volumes (actually, 1916–1920).
[b]To avoid the anomalous situation at the end World War II, the last time interval was increased from ten to fifteen years.

the original tradition of attributing the results of psychological research to individual subjects undergoes a progressive decline. Toward the end of this period most psychological investigations no longer report their results in this way but base their knowledge claims on their use of groups of subjects. In general, the old Wundtian style continues to characterize the areas of psychological research in which it was first introduced, notably the areas of sensation and perception. But in the many new areas of psychological research, both basic and "applied," the new way of establishing psychological knowledge quickly becomes the preferred way.

But this is only a first step in summarizing the historical trend because the categories used here are global in character. Both the category of "individual data" and the category of "group data" are very broad, and cover a number of more specific subcategories. For instance, the "individual data" category could include studies that were truly experimental and presented quantitative data as well as studies that lacked one or both of these features. Actually, this is not a serious problem for this section of the research literature at this particular period because most individual data studies are in fact experimental and quantitative in character. The

Table 5.4. *Percentage of empirical studies reporting individual or group data: Two journals of applied research*

Type of data	Journal of Applied Psychology[a]	Journal of Educational Psychology
1914–1916		
Individual data only	15	11
Individual and group data	7	14
Group data only	77	75
1924–1926		
Individual data only	11	5
Individual and group data	6	4
Group data only	83	91
1934–1936		
Individual data only	5	3
Individual and group data	2	3
Group data only	93	94

[a]The *Journal of Applied Psychology* only began to appear in 1917; the 1914–1916 analysis actually covers volumes published between 1917 and 1919.

heterogeneity of the "group data" category is much more serious. As we have seen, this category might include such diverse groups as those constituted by criteria like age and sex on the one hand and criteria like mental-test performance on the other. Because other forms of group data have not yet been discussed, we need to extend our analysis to a categorization of the different kinds of groups to which psychological data were attributed. This will provide information on more specific trends in the development of investigative practice. Such information is required for a valid assessment of the historical significance of the global trend.

Further analysis along these lines is also needed for the development of a more differentiated perspective on the relationship between the applied and the basic research literature. In general terms it is certainly true that the "applied" literature anticipates developments in the basic research literature as far as this aspect of investigative practice is concerned. In this respect the terms "applied" and "basic" are particularly misleading if they are taken to imply some historical priority of "basic" research whose results are subsequently "applied" in the field. The truth is that psychological research intended for direct practical application developed its own investigative style, which was very different from what had been the traditional style of basic laboratory research.[31] If anything, the lines of influence seem to have operated in the opposite direction. Historically, it was the investigative practice reported in journals of basic research that came to resemble "applied" research in certain respects, rather than the other way around. We will return to this issue in chapters 7 and 8.

Types of collective subjects

The collective subjects that increasingly became the carriers of psychological knowledge claims were obviously not all of the same kind. Fundamentally different defining characteristics can be used to constitute a collectivity. In this section we will consider these defining characteristics more systematically and trace the use of the different kinds of collective subjects that they constitute in the psychological research literature of the early twentieth century.

In distinguishing among the various kinds of collective subjects that have featured in psychological knowledge claims, the most obvious division to make is that between those collectivities that are the result of the investigator's intervention and those that exist irrespective of this intervention. Laboratory science generally constructs its own artificial objects which do not exist in nature, like pure chemical elements and compounds, whereas field science works with naturally occurring objects. An analogous distinction can be made in connection with psychological collectivities. Some are the artificial products of research practice, whereas others exist anyway. Examples of the latter would be age and sex groups, while the former would include the familiar experimental treatment groups. However, it must be emphasized at once that the "natural" groups of psychological research, in the period being examined here, were not usually natural in quite the same sense as the objects of nonhuman field sciences. The "natural" groups of psychological research generally represented social categories that were of great importance in everyday life. They were a salient part of social life outside the laboratory. As a result, psychological research on groups representing such categories tended to have rather direct social implications.

The natural groups of psychological research were taken to represent categories that were important in the organization of social life outside the laboratory. For example, a study that attributed certain psychological characteristics to a group of children defined by their age would have direct implications within a social order marked by an age-graded system of compulsory formal education. Groups representing male and female categories would have an analogous social significance. Psychologists did not create these categories; they simply took them over from the social order of which they were a part. Their research was simply designed to add further information or to specify more accurately the attributes of categories that were accepted without question. The collective subjects of this kind of psychological research were not defined by psychological considerations but by cultural givens.

A very different situation existed where psychologists constructed the collective subjects of their research in terms of their own criteria. In this case the population about whom the research process provided specific information was not related to any particular social category but was simply a product of the researchers' intervention. The earliest examples of this occur when the experimental responses of a number of subjects are averaged to yield a group mean. Innocent though such a procedure seems at first sight, it implies a fundamental conceptual change. Such a group mean is obviously not the attribute of any of the actual persons who contributed to its constitution but the attribute of a collectivity. But what kind of collectivity is this? It is the group of individuals who happened to participate as sources of data in a particular psychological investigation. Their common activity in the experimental situation defines them as a group. There never was such a group before the investigation took place. They do not represent any preexisting social category. If this group represents anything at all, other than itself, it is a hypothetical population of individuals who could all be subjected to the same investigative procedures. In other words, the collectivity of which the group mean is the attribute is one that is defined by laboratory practice rather than by social practice outside the psychological laboratory.[32]

It is difficult to overemphasize the potential importance of this shift for psychology as a discipline. It points the way to a science that supplies its own categories for classifying people and is not dependent on the unreflected categories of everyday life. This has strong implications both for the internal development of the discipline and for its social impact. As regards the latter, there is now the possibility that psychologically constituted collectivities will compete with and even replace some of the more traditional categories of social grouping. Categorizing children or recruits by intelligence quotient was an early example of such a process.[33] In terms of the internal development of the discipline, the artificial constitution of collectivities by the research process itself provided the basis for a science of psychological abstractions that need never be considered in the context of any actual individual personality or social group.

There are three major ways in which the psychologist's intervention produced new groupings of individuals that did not correspond to groupings on the basis of traditional social criteria. The first way occurs when a number of individuals are subjected to analogous experimental treatment and their responses are pooled. A learning curve based on the average responses of a number of subjects would be an example of this process. The relationship between some response measure and the number of trials, which such a curve illustrates, is attributed to a group of subjects and is not expected to be empirically demonstrable in every (or perhaps any) individual member of the group. Such a group is purely an artifact of experimental practice. What defines membership in such a group is common exposure to a particular pattern of experimental intervention. Pre-

senting the average results for the group instead of the results for each
individual suggests that the attributed function is like a general natural
scientific law that holds across individual subjects.[34]

A second kind of artificial group is created when the individual subjects
in an experiment are deliberately subdivided on the basis of their differential exposure to different experimental conditions. Whereas in the first
case the responses of all the subjects who participated in the experiment
are combined after the event, we now get a preplanned division of the
subjects into two or more groups, each exposed to a different pattern of
experimental treatment. What is now of interest is not so much the actual
pattern of group response as the differences between these patterns in
different groups. Such differences can then be causally attributed to differences in the experimental treatment of the groups. In other words, we
are dealing with the kinds of groups that are the product of what has come
to be known as control-group design. Here the groups to which experimental results are attributed are defined by the differential experimental
treatment to which their respective members are exposed. We may therefore call them *treatment groups* to distinguish them from the *general experimental groups* discussed in the previous paragraph. Both kinds of
groups are artifacts of experimental practice, but they must be distinguished
from each other because they have a different history, as we shall see, and
because they imply a different approach to experimentation.

A third kind of artificial group produced by psychological investigation
is the *psychometric group*. In this case the defining characteristic of group
membership is based on performance on some psychological measuring or
assessment instrument. In the historical period examined here this is usually
a psychological test, most often an intelligence test, but it could equally
well be an attitude scale or personality questionnaire. Groups defined by
certain levels of performance on such instruments are obviously the product
of the psychologist's intervention and are not "natural" groups based on
some preexisting social classification. Psychometric groups differ from the
other artificial collectivities we have discussed in the different interpretation
that is given to the psychologist's intervention. In the case of general
experimental and treatment groups, the intervention that defined the group
was thought of as a modifying intervention, but in the case of psychometric
groups the defining intervention was supposed to be one that elicited a
stable characteristic of group members.

Given that in any investigative situation the responses of psychological
"subjects" will depend both on what they bring to the situation and on
what the situation brings to them, all three kinds of artificial groups are
artificial in two senses. The first sense is the one we have already discussed;
it expresses the fact that such groups are created in the process of investigation and do not exist in the everyday world outside this process. But
in creating such groups, psychologists also made them artificial in a second

sense. If we regard the noncorrespondence of these groups to preexisting social categories as a kind of sociological artificiality, then we also should take notice of a fundamental artificiality on the psychological level. The existence of experimental and treatment groups was defined solely in terms of certain modifying factors in the investigative situation, and more or less stable characteristics of individuals were relegated to the category of "error." Psychometric groups, however, were defined solely in terms of supposedly stable characteristics of individuals, and situational effects were either ignored or relegated to artifactual status. Thus, in every case the defining characteristic that constituted the investigated group was based on a psychological abstraction or idealization. In the one case the idealization was that of a collective organism that exhibited only modifiability; in the other case it was that of a collective organism that exhibited only stable traits.[35]

This distinction was important for the kinds of knowledge claims that psychologists were interested in making. In order to make universalistic knowledge claims psychologists took to presenting their data as the attributes of collective rather than individual subjects. Very frequently, these collectivities were constructed by psychologists for this specific purpose, and in constructing them they postulated the existence of a collective organism that already exhibited the assumed general characteristics on which their knowledge claims depended. Thus, to demonstrate the effects of supposedly stable characteristics, they constructed experimental groups defined by such assumed characteristics, and to demonstrate the modifying effects of experimental intervention they constructed groups entirely defined by exposure to such intervention. Such a procedure was inevitable if knowledge claims were to be made through the medium of a myriad of separate, relatively small-scale empirical studies. Problems only arose when the hypothetical nature of the collective subjects that featured in psychological investigations was forgotten, and attempts were made to transfer the results of such investigations to the world of real subjects.[36]

6

Identifying the subject in psychological research

Human sources for psychological generalizations

Making psychological knowledge claims, we saw in the last chapter, involves an ordered set of attributions attached to a particular kind of subject. The significance of these attributions depends in a crucial way on the nature of the subject whose attributes they are claimed to be. If the subject is an individual consciousness, we get a very different kind of psychology than if the subject is a population of organisms. The gradual predominance of the latter type of subject was seen to depend on the interest of psychologists in a type of knowledge that was applicable to populations outside the laboratory but that was as abstract as the generalizations of the physical sciences. This meant working with artificially defined collective subjects in the research situation and then imposing these definitions on people outside the laboratory.

In other words, psychological research on populations had a tendency to replace the social categories that defined populations in real life with populations defined in terms of nonsocial categories. American psychology aimed to be a socially relevant science, but not a social science. Its approach was to be that of a natural science, although its ultimate field of application was to be found among members of real societies.[1] Thus, research on socially defined populations came to be regarded as "applied" research, whereas "basic" research worked with abstract populations. In the extreme case the populations need not even be human. In fact, white rats fulfilled the role of a population of completely abstract organisms more adequately than any human population could hope to do.[2]

But where human populations were retained, the making of appropriately general knowledge claims faced a problem: Each empirical investigation was a historically unique event, involving the interaction of particular participants at a certain place and time, yet some product of this interaction had to be presented as valid independently of these special historical conditions.[3] On one level a psychological experiment is just another social situation in which specific individuals interact at a certain time and place and under particular circumstances. To advance even a mildly convincing claim that, unlike ordinary social situations, this situation yielded some special kind of knowledge, considered ahistorically valid and transsituationally relevant, one has to be able to point to special features of the experimental situation. These special features might be sought in the nature of the participants, the nature of their activity, and the nature of the circumstances to which they were exposed.

To start with the last of these possibilities, one could try to control the circumstances by reducing their apparent complexity, ambiguity, irregularity, or meaningfulness. Instead of having the participants interact under normal life circumstances, one could expose them to pure tones, reduced visual fields, regular stimulus series, meaningless syllables, and so on. In the "results" section of one's research report one could then present these special features of the experimental situation as though they were the only features of the situation that were relevant to one's knowledge claim. It seems that in the early days of experimental psychology claims to scientificity rested very largely on the factor of stimulus control. This could lead to somewhat ludicrous situations, because the reliance on this factor was easily transferred to the (brass) instruments that were the means used to actualize it. Thus we find that early attempts by American experimentalists to enter the field of clinical research were bedeviled by their misplaced attachment to their hardware.[4]

The second special feature of the experimental situation involved the nature of the activity to which the participants were limited. A fixed structure was imposed on the activity of experimental participants. That structure segmented their activity into artificial units that could be counted and aggregated, thus leading to the production of quantitative data that permitted a precise comparison of various effects.[5]

Discussions that have either extolled the merits of the experimental method or criticized it have generally revolved around one or both of these special features. Because psychological experimentation in principle shares these features with experimentation in the natural sciences, such discussions tend to reinforce the view that doing psychological experiments means practicing natural science. But in the reporting of psychological experiments something else happens, namely, the attribution of the data to a human data source. The nature of this source has as much to do with the kind of knowledge claim made in the experimental report as have the other two factors. Although the pervasive belief in a natural-science model of psychological experimentation has generally prevented any discussion of this aspect, it is of profound importance for an understanding of the historical development of psychological investigation as social practice.

Consider the difference between a literary report on a human interaction and its outcome and the standard way of reporting when that interaction is a psychological experiment. One crucial difference between these two kinds of reports is to be found in the way the participants in the interaction are referred to. In the literary report they are likely to be identified by highly diversified individual and historical information, but in the research report any reference to the participants will be stereotyped and relatively poor in information. The claim that the outcome of the experimental interaction is not to be taken as a historically unique product but as a "finding" of potentially universal validity depends as much on a special way of describing and constituting the human data source as it does on other features of the experimental situation.

We therefore need to distinguish between the *identity* of the research subjects and their *identification* in the experimental report. All human participants in experimental situations necessarily have some identity as individuals and as members of certain social categories. They will also be referred to in some way in the published presentation and discussion of the research results. This experimental identification may or may not coincide with their normal personal or social identity. In this section of experimental reports, what is omitted may be just as significant as what is mentioned explicitly. Readily available information about the identity of participants may be omitted from the results section of the experimental reports because it is considered to be irrelevant to the scientific status of the knowledge product.

Another necessary conceptual distinction is that between the identity that participants have outside the experimental situation and the identity they have within it. As noted, the former involves the unique personal identity of the individual and the identity conferred by membership of various social categories like those of age, profession, and educational status. But because experimental situations involve a division of functions, the published reports are also likely to identify the participants in terms of these functions, as either experimenters or subjects. These temporary identities, which begin and end with the experimental situation, are deliberately created in order to construct such a situation.

These distinctions lead to two sets of questions. First, we will try to establish what kinds of extraexperimental identities were most common among those who participated in psychological research situations and ask what significance the various identities had for the knowledge claims that emerged. Second, we will examine the role of intraexperimental identities in the construction of research reports and their knowledge claims.

Social identity of research participants

Broadly speaking, there are two kinds of functions in the experimental situation for which the external social identity of the participants may be highly relevant. The first such function is that of the subject or data source, where the participant's social identity may be relevant because the goals of the investigation treat him or her as a representative of a particular socially defined group, such as male, female, or grade-three child. But there is also the function of scientific observation, reporting and recording of what was observed, which traditionally requires someone with an appropriate social identity. Experimental reports emanating from unqualified persons will not be taken seriously and will not be published in the recognized scientific literature. By the time experimental psychology makes its historical appearance the social identity of those functioning as experimenters and/or authors of experimental reports is on the whole unproblematical within rather narrow limits. The one who vouches for the reliability and accuracy of the observations and for the care with which they have been made must possess appropriate academic qualifications and experience. In other words, he or she must have a particular social identity.

The function of making observations that count as scientific is one that is common to the experimental situation in psychology and in other areas of experimental natural science. The function of the human data source is peculiar to psychology experiments and links them to the social sciences. But there is no reason *in principle* why these two functions should not be carried out by the same person. There may be practical difficulties in combining the two roles and there may be specific reasons for separating them in particular investigations, but in fact many of the classical pioneering studies in experimental psychology combined them very successfully. One need only think of Fechner's work in psychophysics and Ebbinghaus's original work on memory to appreciate this fact. As we saw in chapter 2, the reasons that prompted the early experimentalists to initiate a division of labor in the psychological laboratory appear to have been of an essentially practical nature.

Thus it is entirely understandable that in a sample of three volumes of Wundt's journal *Philosophische Studien* for the period 1894–1896 about 70 percent of the published experimental studies feature academic psychologists in the role of subjects. (In the successor journal *Psychologische Studien,* this proportion even reaches 80 percent for the 1909–1911 period.) In most of these studies, the subject was required to make careful sensory discriminations in the presence of systematically varied external stimuli, and his social status was a guarantee that these difficult and tiring observations would have been made with due care. In this respect experimental psychology was no different from experimental physics or other natural sciences that insisted that only observations made by persons of appropriate background and experience had scientific credibility. The difference lay in the way in which these observational data were *interpreted:* In the case of physics they were interpreted in terms of events in the observed object; in the case of psychology, in terms of events in the observing subject.

For this model of psychological experimentation, prominent in the early years of the discipline, the social identity of experimental subjects was important in the same way that social identity was important in experimental science in general – namely, as a guarantee of the credibility of the observational data. But as we saw in chapter 4, there were other models of psychological investigation extant at an early stage, and for these the social identity of the participants was affected by additional considerations.

One departure from the Leipzig model involved the investigation of psychological processes in representatives of populations with a social identity that was felt to be psychologically interesting. This kind of investigation was in the tradition of clinical experimentation described in chapter 4. Early examples involved categories of persons with special abilities or deficiencies – somnambulists, blind or intellectually deficient persons as well as musically gifted persons and lightning calculators. In all these cases the individual functioned as a subject of psychological investigation much as someone with an unusual physiological or pathological condition might function as a subject for medical investigation. The difference was that whereas medical science already selected its cases by imposing its own categories, early experimental psychology essentially adopted culturally established categories and treated them as psychological (or biological) categories. In this model of psychological research, therefore, the social identity of individuals in the subject role was the crucial factor that pushed them into that role and kept it separate from the investigator's role. Being an appropriate candidate for medical or psychological investigation went with membership in certain social categories. In these cases the role that the subject played in the experimental situation was essentially an extension or elaboration of the role that went with his or her social identity.

Another early departure from the scientific norm of the psychologist–subject–observer occurred in studies that amounted to a kind of psychological census taking.[6] Such studies were essentially descriptive. Mental phenomena were not studied as processes but as isolated instances to be counted. They were not related to any context, and the individuals who reported them only acted as a kind of neutral medium through which these phenomena manifested themselves. In their recording of the incidence of various psychological phenomena, these investigations seemed to be modeled on quantitative natural history. Because the phenomena to be studied were not treated as processes to be analyzed, no special observational skill or experience was required from the data source. Thus, virtually any unimpaired adult with a certain level of literacy could act as a subject for

Table 6.1. *Background of those participating as subjects in published psychological studies*

	Academic psychologists	Undergraduates	School children	Noneducational
1894–1896	52	38	13	3
1909–1912	24	38	28	9
1924–1926	11	40	36	11
1934–1936	10	52	27	14

Note: See the appendix for notes on this table.

these investigations. The only requirement was that they be available in some numbers to permit a convincing counting of instances. Undergraduate students fitted this role perfectly.

Research students also appeared as experimental subjects in some numbers. However, they functioned as expert observers, exactly like fully qualified academic psychologists. Their use, therefore, introduces no new considerations into the present analysis.

Children, who constitute the last of the major categories of subjects for psychological research, do need special consideration. At least from the age of school entry onward, children were always a significant source of psychological data for American investigators. In the early years school children were prominent subjects for the practice of psychological research as a kind of quantitative natural history. Much of this work was inspired by G. Stanley Hall, as we have already noted. School children shared with undergraduates the advantage of an institutional context that made them conveniently available in socially classified groups and compelled their participation in the research process. But, unlike undergraduates, school children were also of great interest as potential objects of practical psychological intervention. They constituted by far the most important category of persons with special deficiencies and were therefore appropriate objects for clinical experimentation.

Table 6.1 presents some temporal trends in the use of different categories of research subjects. Certain general trends emerge rather clearly. Academic psychologists are at first the most important group of subjects for psychological research and then show a progressive decline. This is undoubtedly a reflection of the fate that befell the Wundtian model of psychological research practice, as noted in chapter 4. In view of this decline it is perhaps surprising to find that the use of subjects from outside educational institutions increased only very gradually during the period covered. Moreover, during the 1920s and 1930s most of the studies employing this category of subject come from a single source, the *Journal of Applied Psychology*. The core of the discipline remained uninterested in most of the human population as a source of research subjects. When experimental psychologists ceased to take the subject role themselves they turned to animals, to children, or to undergraduates as the three alternatives. Among these, undergraduates were clearly the favorites. They had always been a major source of data in American psychological research, and in the period after World War I they reach a position of predominance which they have never lost.[7]

There was also a change in the way in which students were used. In the early period it was quite unusual to publish studies in which undergraduates were the *only* source of experimental subjects. As a rule, undergraduates would only appear in studies that had also used the services of other kinds of subjects. Comparisons of results from subjects with different backgrounds were not uncommon. This continued to be the predominant pattern up to the second decade of the twentieth century. However, in the period after World War I studies based only on undergraduate subjects became much more frequent and outnumbered those studies in which both undergraduates and other categories of subject were employed. By the 1930s, two-thirds of the studies that used undergraduates used them as the sole source of psychological data. The role of undergraduates as research subjects seems to have changed. At first, their extraexperimental identity still had some significance for their research role, insofar as their reactions were compared with those of subjects with a different identity. But, increasingly, they came to be used as a convenient source of abstract experimental subjects whose extraexperimental identity was considered to be irrelevant in the context of psychological experimentation.

Even in the days when it was common for subjects with a different extraexperimental identity to be compared with each other, the investigators did not usually think in terms of social identities. The categories in which most early American psychologists thought about the objects of their research were derived from biology, not from sociology or history. So the comparison of different kinds of subjects was more likely to be interpreted in terms of quasi-biological categories, like age or innate endowment, than in terms of history or culture. Gradually, the link to biological categories became more and more remote, as psychology developed its own abstract categories for the description of human action. The further this process developed the more completely substitutable did experimental subjects become. Individuals simply became the media through which abstract laws of behavior expressed themselves. Because the abstract laws and not their individual carriers were the primary objects of investigation, the social and even the biological identities of experimental subjects became theoretically irrelevant. The choice of subjects could be governed solely by practical convenience. On that criterion no group of human subjects could compete with college students.

The rhetoric of experimental identities

Published research reports reflect certain decisions that were taken about how to refer to the human source of the data. The author of such a report has the alternative of referring to such sources either by giving them their ordinary social identities, such as boys or girls, or by giving them special experimental identities, such as subjects or observers. This choice is not made randomly but depends on prevailing norms and conventions among the community of investigators. It is likely that these norms reflect prevailing conceptions of psychological knowledge.

One might think that the choice of how one refers to the human sources of one's data is a relatively trivial matter. This appearance of triviality is however the product of a situation in which the norms governing such matters have simply become hallowed by long usage, so that the awareness of their implications has been lost. When we look at the historical development of these norms we come across issues that are far from trivial.

In the first place, the history of experimental psychology is marked by a long period of indecision about the appropriate way to refer to the experimental identity of those who functioned as the source of data. In the very early days of the discipline a number of terms were in use (see chapter 2), but in the English-language literature two terms quickly overshadowed all the others: "subject" and "observer." By the end of the nineteenth century the former was being used in about half of all published reports, and the latter in about a quarter. But then it took another half-century for the use of observer to dwindle to a negligible level. Clearly, the connotations of these terms were not a matter of indifference at the time but presented authors and editors with a real choice.

In Table 6.2 the identification of the human data source as either subject or observer is traced over four time periods. Although a few studies employ both terms, most opt for one or the other. In general, the term subject has an early preponderance, which increases very gradually over the period under review. It is as though the rival term observer were fighting a determined rearguard action. The main center of resistance appears to be concentrated around the *American Journal of Psychology*. But this is not apparent until after the turn of the century. Between 1895 and 1910 a major swing from subject to observer takes place in this journal whose countertrend remains visible until the 1930s. It seems most likely that the original switch to the term observer in preference to subject was linked to the emergence of a more self-conscious kind of introspectionism, which presented itself as "systematic experimental introspection" (see chapter 3). In America, Titchener was the leading figure in this development, and the *American Journal of Psychology* became its main publication channel. This defined the role of the psychological data source entirely in terms of

Table 6.2. *Percentage of empirical studies identifying the human data source as "subject" and as "observer"*

	Subject				Observer			
	1894–1896	1909–1911	1924–1926	1934–1936	1894–1896	1909–1911	1924–1926	1934–1936
American Journal of Psychology	69	25	49	55	38	72	56	45
Psychological Review[a]	68	75			26	33		
Psychological Monographs	—[b]	82	61	73	—	29	43	8
Journal of Experimental Psychology	—	—	92	88	—	—	12	20
Journal of Applied Psychology	—	—	71	35	—	—	0	0
Journal of Educational Psychology[c]	—	36	42	46	—	12	0	0

[a] Coding discontinued for third and fourth period because of paucity of empirical studies.
[b] Dashes indicate periods before journal commenced publication.
[c] Publication began 1910; hence first three volumes coded are 1910–1912.

introspective observation, whereas earlier practice had been much less restrictive. Not surprisingly, it also favored research areas whose content gave the most scope for the working out of its methodological program.

It seems that as a result of this development, and the subsequent reaction of the behaviorists, subject and observer acquired some ideological significance in terms of the internal divisions within American psychology. In the 1920s these terms were being pressed into service by rival schools of experimental practice. By 1929 one well-known experimental psychologist[8] considered the time to be ripe to urge the dropping of the term observer from psychological discourse. After all, "in many contemporary lines of psychological investigation the so-called 'observer' does no observing." This led to a prompt reply (in the *American Journal of Psychology,* of course) by a prominent representative of a different point of view,[9] who, while no longer defending Titchenerian introspectionism, objects to a "cult of objectivism" that refuses to recognize the special nature of those experiments "where the organism enters the scene as *agent"* (emphasis in the original). Further controversy[10] did not resolve the divergence of perspectives.

Although the ideological connotations of the terms subject and observer were probably salient only for a minority of experimentalists, the failure of the discipline to establish uniformity of usage even after several decades is a reflection of a continuing division of opinion about the essential function of the human data source. There appears to have been a general feeling that the data source ought to be identified by reference to its essential function in the experimental situation. But what was the essential function? The "objectivists" clearly felt that it consisted in providing material for the experimenter's observation and use, and the term subject had by long usage acquired this kind of meaning (the medical prehistory of the term had been forgotten, although it was preserved in practice). The minority who resisted this trend saw the essential function of the human data source somewhere else, namely, in its competence to comprehend the experimental task and act appropriately. This latter interpretation was of course quite close to Wundt's original conception of the role of the human data source in psychological experiments. As we have noted, experimental psychology began life with several different conceptions of its practice. One of these was in retreat almost from the start, although complete uniformity of practice does not appear to have been reached during the period under review here.

Both subject and observer refer to a strictly intraexperimental identity. Persons only have these identities insofar as they play a particular role within the social situation constituting a psychological investigation. However, these same persons do not thereby lose their ordinary social identities and may still be identified by the latter in experimental reports. The frequency of such "extraexperimental" identifications in published empirical

Table 6.3. *Percentage of empirical studies identifying the human data source by an extraexperimental identity*

	1894–1896	1909–1911	1924–1926	1934–1936
American Journal of Psychology	19	11	5	20
Psychological Review[a]	13	18		
Psychological Monographs	—[b]	6	46	56
Journal of Experimental Psychology	—	—	28	7
Journal of Educational Psychology[c]	—	60	62	63
Journal of Applied Psychology	—	—	33	65

[a]Coding discontinued for third and fourth period because of paucity of empirical studies.
[b]Dashes indicate periods before journal commenced publication.
[c]Publication began 1910; hence first three volumes coded are 1910–1912.

studies is shown in Table 6.3. Most commonly these extraexperimental identifications refer to "children" and "pupils," less commonly to "boys" and "girls," and occasionally to "students" and to "men" and women."

A clear difference emerges between journals devoted to "basic" experimental research (*American Journal of Psychology* and *Journal of Experimental Psychology*) and those more concerned with "applied" knowledge (*Journal of Educational Psychology* and *Journal of Applied Psychology*).[11] Subjects of psychological research were, on the whole, more likely to be identified by an extraexperimental identity if the investigation was intended to have some direct relevance outside the investigative context. In experimental work devoted to abstract psychological knowledge, however, the human source of the data was very likely to be identified only by an intraexperimental identity.

It is also the case that studies published in the "applied" journals were, on the whole, more likely to be based on the use of subjects with a non-academic background, either children or adults. In the *Journal of Applied Psychology* and the *Journal of Educational Psychology* the proportion of studies employing such subjects ranged between 50 and 95 percent for different periods, but in the *American Journal of Psychology* and the *Journal of Experimental Psychology* the range was between 11 and 37 percent. The research published in the latter two journals was more likely to be based on student subjects or, in the early period, on psychologists acting as subjects.

In general, the research literature that is based on the preferential use of undergraduates is most likely to identify its data source by an intraexperimental identity, whereas that which is based on the use of children or nonacademic adults is more likely to retain extraexperimental identities in reporting research results. An overall comparison for all six journals and all time periods covered in Tables 6.1, 6.2, and 6.3 indicated that children were five times more likely to be referred to by their common extraexperimental identities than were college students.

Why were children and nonacademic adults much more likely to retain their normal identity than were college students when they participated in psychological investigations? The conferral of particular identities in social situations depends on the perceived relevance of those identities for the social situation. There is no reason to suppose that the social situation that we know as a psychological experiment will be any different from other social situations in this respect. Whatever identities are conferred on the participants in the experimental situation will be the ones that are most relevant to the construal of that situation by those who report on it. If children often retained their extraexperimental identity in the experimentation situation whereas undergraduates were usually given a special identity, this must mean that the child's ordinary identity was considered to be experimentally relevant in a way that the undergraduate's was not.

In other words, investigations involving children were usually regarded as providing information specifically about children, whereas most investigations involving college students or academics were not regarded as providing information specific to the psychology of college students or academics. Sometimes the educational status of the college students who served as subjects was considered relevant to the purpose of the research, and in these cases we do find them referred to as "students."[12]

Conversely, the tendency to refer also to children as "subjects" increased steadily during the period covered. This may have been part of a more general tendency to emphasize the scientific pretensions of the discipline, pretensions that demanded that psychological research be concerned with abstract relationships rather than with people.

Experimental psychologists were concerned to establish a kind of knowledge about human beings that would be ahistorical and universal. Yet to obtain this knowledge they had to work with specific, historically defined human data sources, and they had to extract their data from these sources in investigative situations that were also historically specific. To cope with this paradox, psychologists commonly employed certain rhetorical devices when reporting their data. It became customary to emphasize the experimental identity of human data sources at the expense of their ordinary personal and social identity. Thus, their actions, as reflected in the "experimental results," were not attributed to them as historical individuals but to them as incumbents of a special experimental role. Where experimental results are presented as activity attributed to persons who have no identity apart from their identity as experimental subjects, the implication is that no other identity is relevant in the experimental context. The experiment and its results stand apart from the sociohistorical context in which the rest of human life is embedded. It appears to lead straight to human universals, not infrequently identified with biological universals.

A further development that enhances this effect involves not attributing the experimental results to specific individuals at all, not even individuals that only have an experimental identity. Experimental results are then presented as the attributes of experimental groups whose anonymous members have no individual existence in the experimental report. This practice, discussed in chapter 5, can now be seen as having the additional function of enhancing the isolation of laboratory products from the personal and cultural reality that produced them.

The fact that the handling of the attribution problem in the presentation of experimental results involves rhetorical considerations does not mean that the knowledge claims implied in this presentation are necessarily false. It only means that they have not been empirically secured, that the issue remains wide open. Published reports of experimental results cannot avoid the task of establishing the claim that what is being reported can justifiably count as "scientific" information. For a discipline that took "scientific" to imply reference to some universal truth beyond individuality, history, and local meanings, the establishment of claims to being scientific frequently depended on appropriate manipulation of the identity of the sources to which data were attributed. With few exceptions, claims to universality were not grounded empirically but were established by fiat. The role of the experimental report was, among other things, to create the illusion of empiricism. By the way the experimental results were presented, the illusion was created that they were not really the product of a social interaction among certain human personalities in historical time, but that they were the manifestation of abstract transpersonal and transhistorical processes. As experimental psychology progressed, its published reports dealt more and more with abstractions and became less like accounts of particular investigations.

7

Marketable methods

Education as psychology's primary market

Thus far, we have traced the development of some of psychology's investigative practices in terms of the general knowledge interests pursued by psychologists. This may be adequate for the more general features of these practices but when we turn to more specific features it becomes necessary to take into account a broader social context. The fact is that almost from the beginning of the twentieth century psychology ceased to be a purely academic discipline and began to market its products in the outside world. That meant that the requirements of its potential market were able to influence the direction in which psychology's investigative practices were likely to develop. Practices that were useful in the construction of specific marketable products were likely to receive a boost, whereas practices that lacked this capacity were henceforth placed under a handicap.

Of course, the requirements of the market did not act on a passive discipline. Not only did many psychologists actively court practical application, but these efforts would have led nowhere if the discipline did not have at its disposal certain techniques that lent themselves to the development of a socially useful product. In the present chapter we will analyze two such techniques in terms of their significance for a socially relevant investigative practice: the Galtonian approach to individual differences and the experimental use of treatment groups. Both of these methodological innovations assumed enormous importance in the subsequent development of modern psychology, but it is doubtful whether this would have been the case had they not played a crucial role in the constitution of psychology as a socially relevant discipline.

However, if there is to be any real analysis of the social significance of these techniques, the category of social relevance will have to be brought

101

down to earth. The abstract ideal of a socially useful science may have occupied an important place in the ideology of many American psychologists. But as soon as they tried to convert this ideal into practice, they had to accommodate themselves to the specific opportunities offered by a particular historical context. In principle the possibilities of applying psychological knowledge might be unlimited, but the actual possibilities available here and now were always sharply circumscribed. They depended on existing institutional forms and on the requirements of those who could command the social resources for putting psychological knowledge to work.

Historically, the one area of social practice with which psychology had established firm links was the field of education. During the nineteenth century the Herbartians had laid the groundwork here, and even though their role in the development of modern psychology was minimal, the expectation that psychology had something to contribute to education had been firmly established. When psychology was transformed from being primarily a system of ideas to something more like a system of practices this expectation was naturally extended to those practices. Not only was this extension made by some psychologists, but it was not an altogether alien suggestion for many educationists. However, the specific nature of the contributions made by modern psychology would depend on the fit between what the psychologists could deliver and what the educationists required.[1]

What was it that the educationists required? Given the enormous differences in educational institutions and in the social demands placed on these institutions in different countries, there was considerable national divergence. In the present chapter we will focus on the situation in the United States where the collaboration between educationists and psychologists was particularly effective in the early part of the twentieth century. However, the term "educationists" is almost as vague as the term "social relevance." There were all kinds of educationists, from humble grade school teachers to powerful superintendents of school districts, and their requirements were far from identical.

During the period when the involvement with educational issues played an important role for American psychology there appears to have been a definite shift in the pattern of professional alliances. In the earlier period, characterized by the child study movement of the 1890s, links between psychologists like G. S. Hall and teachers were clearly prominent.[2] The kinds of investigative practices in which this collaboration was expressed were of the crude, psychological census-taking variety referred to in chapter 5. Nothing more sophisticated would have worked, given the limited training and scientific background of the teachers. This alliance may have done nothing to enhance psychology's status as a science, but it probably helped to popularize the subject and, above all, it translated a vague conceptual link between psychology and education into a tie between two professionalizing groups.

After the turn of the century psychologists' relations with teachers became increasingly overshadowed by a new professional alliance which was consummated through the medium of a new set of investigative practices. The group of educators with whom psychologists now began to establish an important and mutually beneficial alliance consisted of a new generation of professional educational administrators. This group increasingly took control of a process of educational rationalization that adapted education to the changed social order of corporate industrialism.[3] The interests of the new breed of educational administrator had little in common with those of the classroom teacher. Not only were the administrators not directly concerned with the process of classroom teaching, they were actually determined to separate their professional concerns as much as possible from those of the lowly army of frontline teachers.[4] In this context they emphasized scientific research as a basis for the rationalized educational system of which they were the chief architects. In the United States the needs of educational administration provided the first significant external market for the products of psychological research in the years immediately preceding World War I.

But to the potential clients of the academic psychologists research meant something that was rather different from the latter's traditional laboratory practice. Research, to the administrators, was an activity whose results had to be relevant to managerial concerns. It had to provide data that were useful in making immediate decisions in restricted administrative contexts. This meant research that yielded comparable quantitative data on the performance of large numbers of individuals under restricted conditions. Excluded was research that went beyond the given human and social parameters within which the administrators had to make their decisions. It was, in other words, technological research that would help in dealing with circumscribed problems defined by currently unquestioned social conditions. Not infrequently, administrators simply needed research for public relations purposes, to justify practices and decisions they judged to be expedient.

In representing the problematic that defined their goals these early educational administrators very frequently made use of an industrial metaphor. As one of them put it in an unforgettable image, sustained over several pages, "education is a shaping process as much as the manufacture of steel rails."[5] Elaborate analogies were drawn between stockholders and parents, general managers of factories and superintendents of schools, foremen and principals as well as industrial workers and teachers.[6] Within this framework nothing was more natural than the extension of the principles of "scientific management" from industry to education. In the considerable literature devoted to this topic, the most succinct formulation of these principles, as they were understood by the educational administrators, appears to have been given by Frank Spaulding in 1913:[7]

1. The measurement and comparison of comparable results.
2. The analysis and comparison of the conditions under which given results were secured – especially of the means and time employed in securing given results.
3. The consistent adoption and use of those means that justify themselves most fully by their results, abandoning those that fail so to justify themselves.

Such a scheme assigned an important and clearly defined role to research. It had to provide the information necessary for steps one and two, so that the administrator could take the appropriate action specified by step three. Two kinds of research data were required: first, comparable measurements of results, defined as individual performance; and second, a comparison of the relative efficiency of various conditions in producing these results.[8] What was needed were scales that measured performance and experiments that assessed the relative effectiveness of such conditions as were potentially under the control of the administrators who had to choose between them. The primary consumers of the results of such research were those whose controlling position made the results potentially useful to them, not those who participated in the research as subjects.

A significant number of psychologists quickly responded to the requirements of a type of research that was quite different both from the kind of laboratory practice that had so far defined scientific psychology and from the computations of child study. By 1910 this new kind of research practice had gained sufficient momentum to lead to the publication of a separate journal devoted to it, the *Journal of Educational Psychology*. The core of contributors was formed by those, like Thorndike at Columbia Teachers College, whose links with educational administrators had already been firmly cemented.[9] But even fairly traditional experimentalists, like Carl Seashore and Edmund Sanford, found the siren call of the new applied psychology immediately appealing. "So long as the science must be financed by appropriations and endowments," Sanford wrote in the first volume of the new journal, "it [the new applied psychology] furnishes an effective defense against ignorant and hostile criticism and a tangible excuse for the investment of institutional capital."[10]

What Sanford did not anticipate was the fact that the attempt to apply psychology to the requirements of educational administration would have consequences for psychology that were far more profound than any contribution psychology was likely to make to education. In the first place, it meant a decisive break with the kind of educational psychology that had been envisaged by the pioneering giants of American psychology, by James,

Baldwin, Hall, and Dewey. Their broader vision was now replaced by a much narrower and purely instrumental conception of what psychology could accomplish.[11] The institutional constraints that the new educational psychologists took for granted required them to emphasize the passivity of the child and to restrict themselves to measured performance rather than wasting precious resources on an exploration of mental processes that had no obvious utility in terms of the goal of choosing the conditions that were most efficient in producing predetermined results.

The new pattern of psychological research practice that originally crystallized in the work of educational psychologists proved to be increasingly attractive to other psychologists in the years that followed. The experience gained in research for educational administration made it possible for psychologists to exploit the professional opportunities presented by the large-scale requirements of military administration in World War I. Without the methods of mental measurement recently tried out within an educational context, American psychologists would have had nothing immediately useful to offer the military authorities.[12] It was at this time that the entire field of applied psychology came to be defined in terms of psychological knowledge that would be useful in administrative contexts. During the 1920s and 1930s the conviction that psychological research had to produce the kind of knowledge that would be potentially useful in certain practical contexts characterized an important and growing section of academic psychologists. For people like Woodworth, psychology's affiliation with education had replaced its earlier affiliation with philosophy.[13] But what this concern with applicable knowledge generally came down to in practical terms was the desire for knowledge that was marketable in administrative contexts. This necessarily entailed a growing preference for a type of research practice that was capable of yielding the kind of data that could be perceived as relevant to the interests of administrators.

Close links between social administration and psychological practice entailed significant effects for both. Certainly, psychological practice had social repercussions, and at least some aspects of this relationship have received a great deal of attention in the past.[14] But of interest in the present context is the reciprocal influence of social practice on psychological investigation. That such an influence existed was clear to central figures like Thorndike, who was certain that "the science of education can and will itself contribute abundantly to psychology."[15] More particularly, the demands of a certain kind of market greatly favored specific research practices that could produce the kind of knowledge that was usable in this context.

In the broadest terms, the kind of knowledge that was most obviously useful in administrative contexts was statistical knowledge. In such contexts information about persons was generally of interest only insofar as it pertained to the categorization of individuals in terms of group characteristics. Dealing with individuals by categories constitutes the essence of administrative practice. One general consequence of the impact of administrative concerns on psychological research has already been described in chapter 5: Psychologists increasingly turned away from an individual attribution of their findings to an attribution that worked with statistically constituted collective subjects. It is not surprising that this development characterized the applied-research literature from its beginnings. Applied research generally aimed to supply administrators with the kind of knowledge that they would need to make efficient decisions. Only quantitative statistical knowledge could serve this purpose in most practical situations.

But it is possible to go further than this global characterization of the new psychological knowledge products. The requirements of the more articulate administrators were in fact more specific, as we have seen. They wanted methods for the comparison of conditions and the measurement of results. Psychologists obliged by concentrating on certain incipient aspects of their own practice that could be directly adapted to these requirements. The result was that these aspects attained an extraordinary prominence in psychological research practice. What were these aspects?

In a strategic article, published in an early volume of the *Journal of Educational Psychology,* E. C. Sanford, a distinguished researcher who was by this time president of Clark College, identified three basic methods that psychologists would be contributing to education as a rationalized social practice:[16] First, there is standard laboratory work, as exemplified by German investigations of "economic methods of memorizing"; the pedagogical implications of this work had been explored by Meumann, but the basis of the method had been established by Ebbinghaus. Second, there was the work on individual differences as exemplified by the method of mental testing that was just then getting into its stride. Third, there was the classroom experiment, whose major pioneer was W. H. Winch in England; in this type of investigative practice groups of school children, matched in ability, would be exposed to different methods of instruction and their performance before and after the experimental intervention would be assessed. The first and third methods were dedicated to comparing the efficiency of different techniques of learning and instruction; the second method was used to select individuals for programs rather than the other way around. All three methods were nicely in tune with the stated requirements of educational administration.

In examining the kind of investigative practice that was promoted by psychology's reliance on a particular market for its research products, we will begin with the second and third of Sanford's methods, mental tests for individual differences, and classroom experiments. These depended very directly on the educational market. The study of the economy of memorizing had more solid roots in certain practices of laboratory psychology, which will be examined in chapter 9.

Mental testing as an investigative practice

The too-ready marketability of mental tests profoundly affected the shape of psychological research. In the period after World War I there was an enormous expansion of work based on the use of mental tests. This work had so little connection with the more traditional forms of psychological experimentation that it eventually seemed as though there were two distinct disciplines of scientific psychology, one experimental and the other based on the correlational statistics used in mental-test research.[17]

Mental testing flourished because of an interest in individual differences, but this observation hides more than it reveals. The term "individual differences" taken in isolation from a specific context is exceedingly vague and could just as easily apply to the work of the novelist as to that of the psychologist. Clearly, when mental testing is derived from an interest in individual differences, it is not this very general meaning that is relevant but a very specific meaning that is conveyed by the context. It was an interest in looking at individual differences in a particular way that found expression in the development of mental tests.

Indeed, the investigation of individual differences preceded the development of modern mental testing by many years. There were old interpretive practices of reading an individual's character with the help of bodily signs. These might be based on somatic indications, as in the classical doctrine of temperaments, or on facial characteristics, as in the relatively more recent versions of physiognomy.[18] During the first half of the nineteenth century the practice of phrenology built on, and to a large extent replaced, these older practices. Somewhat later, the measurement of the brains of various kinds of people and the physiognomic characterization of criminals[19] followed on in the same tradition. Gradually, measuring practices came to replace the older interpretive practices, reflecting a changing conception of the nature of scientific knowledge. What distinguished mental testing from these older approaches was neither an interest in the assessment of individual differences nor a preference for quantitative methods as such. The difference resided rather in the medium through which individual differences were assessed. Mental tests severed the very ancient link between psyche and soma and proposed to assess individual differences entirely by measuring function. In practice, this meant measuring individual performance in restricted situations.

Assessing individual differences in terms of specific performance measures, while relegating any other method of assessment to the realm of the unscientific, amounted to a redefinition of the object of investigation. Psychological differences were now defined in terms of differences on measures of individual performance. It was performance at uniform set tasks that counted psychologically, not, for example, facial expression or artistic style.

Even on this somewhat restricted basis a broadly conceived study of individual differences was entirely possible. In fact, during the early modern period in the investigation of individual differences most investigators were not interested in the comparison of individual performances as such but in the characterization of psychological types and of human individuality. Before the advent of intelligence tests, Alfred Binet's name was associated with an "individual psychology" that was truly a psychology of individuals, in the sense that psychological performance measures were used to assess the individual style of a person's functioning.[20] Another European authority in this area, William Stern, distinguished between the study of human variety and the study of individuality, and he was inclined to accord a higher value to the latter.[21] James Mark Baldwin had criticized the Wundtian style of psychological experimentation for treating individual differences as a nuisance, rather than as an object of study. But he conceived of individual differences in typological terms, as was general at the time.[22]

What the development of mental testing did was to redefine the problem of individual differences in terms of a *comparison* of performances. The quality of a performance could no longer be used for describing individuality or for analyzing typological patterns. Instead, a measure of individual performance was to be used for specifying the individual's position with respect to an aggregate of individuals. Comparing individual performance to a group norm, however, meant that the characterization of any individual depended just as much on the performance of a set of others as it did on anything he did himself. The whole exercise depended on the assumption that the salient qualities for characterizing an individual were qualities that he or she shared with others rather than qualities that were unique.[23] These common qualities had to be thought of as constant elements whose nature was unaffected by their cohabitation with other such elements in the same individual. Carried to its logical conclusion, this methodology for assessing "individual differences" actually eliminated individuals by reducing them to the abstraction of a collection of points in a set of aggregates.[24]

Although the statistics of individual differences constituted the very antithesis of an interest in psychological individuality, they were able to speak quite directly to another kind of concern: the problem of conformity. The new psychological practice was based on the setting up of "norms" in terms of which individuals could be assessed. In most cases these norms were psychological only by inference; in the first place they were norms of social performance.[25] The categories in terms of which individuals were graded were not generally socially neutral categories but carried a powerful evaluative component. It was not an academic interest in the psychology of human cognition that motivated the normative study of individual performance but an interest in establishing who would most effectively conform to certain socially established criteria. These criteria ranged all the way from the unidimensional "general intelligence" of the eugenicists to the qualities needed in a good salesman. But they were always criteria that only made sense in the context of particular social interests, be they grand and ideological or practical and mundane.

Throughout most of the nineteenth century, the concern with individual differences had been fueled by two different sets of interests, one private, the other public. In a competitive society, offering certain possibilities for individual social mobility, individuals had an interest in assessing their personal potential so as to make good use of any opportunities for social advancement that might present themselves. For several decades phrenologists were able to exploit this as a market in aids to self-improvement,[26] and to some degree psychologists were later able to do likewise. In the course of time, however, a much more important market opened up for experts on individual differences. Those who administered social programs and institutions always had a potential interest in such expertise, insofar as it could assist in the administrative process – specifically by providing a culturally acceptable rationale for the treatment of individuals by categories that bureaucratic structures demanded.

With the enormous increase in the size and scope of bureaucratically administered institutions during the latter part of the nineteenth century, especially in the fields of education and the management of social deviance, the need for special sorting practices increased greatly. In fact, social sorting on apparently rational criteria became the major function of these institutions. At first, the task was accomplished by adapting a social practice that had recently become popular in a related context, namely, the *examination*. Two kinds of examination had existed for a long time, the academic and the medical. By the middle of the nineteenth century both had been adapted to new tasks. The academic examination had been adapted to the selection of senior administrators, and the medical examination had been transformed into the psychiatric examination, which, among other things, assessed moral fitness.[27] In the latter part of the century the academic examination was further modified to make it suitable for the mass selection and grading of ever larger sections of the population, while the distinction between intellectual and moral worth was becoming blurred. The advent of mental tests represented a continuation and culmination of this development. Finally, the notion of a unitary, biologically fixed "intelligence" provided a license for grading the entire population as though they were members of one school class.

Those who were in control of the social institutions in which mental tests found their natural home were looking for ways to sort individuals in terms of their measuring up to institutionally appropriate criteria. They needed effective techniques for making selection decisions that would enhance the efficiency of existing institutional structures. The characterization of individuals in terms of their degree of conformity to criteria of performance within specific social contexts made it easy to redefine the problems of institutions, or even the problems of society, as individual problems. If all social problems were nothing more than the aggregate of individual problems, they could be handled by appropriate treatment of individuals and required no questioning of the social order.

Within psychology the growing popularity of research based on one or other form of mental testing accounted for a considerable part of the expanded use of "collective subjects," which we noted in chapter 5. This involved the large-scale appearance of both "natural" and psychometrically defined groups in psychological research reports. No longer did psychological research mean the experimental exploration of individual minds. The individual was of interest only in terms of his or her standing in an aggregate. Research objectives largely shifted to the comparison of such aggregates and the statistical relationships between them. In certain practical settings ("psychological clinics"), a more individualized employment of mental tests, closer to the original vision of Binet, did continue.[28] But insofar as such practice claimed to have a scientific basis, that basis was also statistical. Even though individual patterns might be considered, they were still patterns of performance defined in terms of common group norms. Attempts to relate clinical practice to an individualized experimental psychology were unsuccessful.[29] What survived were packets of individualized clinical practice whose claims to scientific expertise depended on their reliance on statistical research data. Practically relevant research practice was generally devoted to work with statistically constituted psychological objects.

Marketable mental testing ensured a dominant place in psychological research for the style of investigative practice that had been pioneered by Galton. This applied to the social construction of the investigative situation where some of the features of Galtonian anthropometry were exaggerated to the point of bizarre caricature in the mass-testing methods developed in military, but also in educational, contexts.[30] But it also applied to the way in which statistical techniques were used to create the objects that were the real focus of the researchers' scrutiny. These techniques no longer functioned to make the observations of individuals more precise but came to play a crucial role in constituting the objects around which research revolved – statistical distributions of scores contributed by a mass of individuals.[31]

The ready adoption of the Galtonian model by a significant section of American psychology is not at all surprising when one takes into account certain parallels between the situation faced by Galton and that which confronted early twentieth-century American psychologists. Galton wanted to establish a science of human heredity although it would be impossible to ground such a science in anything resembling controlled experimentation. His alternative was to base this science on the statistical comparison of group attributes.[32] Many American psychologists were interested in promoting a socially relevant science involving aspects of human conduct and performance that were not accessible to precise experimental study under controlled conditions. The emerging science of biometrics offered statistical techniques, initiated by Galton and developed by Pearson, which promised to provide ways around this problem. Something that looked like a science could apparently be created by statistical rather than experimental means.

There was however a residual problem about establishing psychological knowledge claims on a Galtonian basis. In the classical psychological experiment, devoted to the exploration of functional relationships (see chapter 2), there was always a fundamental asymmetry that provided the basis for causal attributions. One of the variables was under the experimenter's control and could be "arbitrarily" (Wundt) varied. If this led to regular changes in the response measure, one had grounds for making causal attributions. The situation was quite different in the Galtonian model. Here the investigator attempted to impute relationships in situations over which he had no control. Such relationships had therefore to be symmetrical, because the investigator was unable to influence the situation so as to give the status of cause to one of the variables and the status of effect to the other. What he could hope for was a measure of concomitance or covariation. The descriptive statistics of the Galton–Pearson school[33] provided him with such measures. Such statistics were therefore constitutive of the relationship described, whereas the calculus of error of the Wundtian model could not be said to constitute the causal relationships that were investigated. In the Galtonian model the establishment of statistical correlations among attributes of natural populations became the goal of research, whereas in the classical model statistics had been at best a means for ensuring the reliability of certain observations.[34]

The adoption of statistical usage in the tradition of Galton and Pearson meant that American psychology was now committed to making two very different kinds of knowledge claim, both of which had serious grounds for being taken as scientific. At first, this produced considerable confusion and a bewildering variety of rationalizations within the discipline.[35] Gradually, however, a kind of official doctrine emerged. This established the ideal form of research practice as one that combined the manipulative aspects of experimental procedure with statistically constituted objects of investigation. It was not noticed that this transformation deprived the experimental manipulation of its original rationale, which was the production of some causal process in an individual psychophysical system. Interest had shifted to the statistical outcome of experimental trials as manifested in differences between individuals.[36]

This was quite acceptable to the growing number of American psychologists who saw their scientific work as devoted to the development of techniques for the social control of individuals. Such a psychology had to be able to use its measurements to make predictions about future effects that could be taken into account by those in controlling positions. In principle such predictions about future effects, reliably to be expected on the basis of previous measurements, could of course be made on a purely statistical basis. But that would have required really large-scale research with a very extended time frame. Psychology would have to be far better established before it could hope to muster the resources for such research. Even then, the social organization of research and of individual career patterns greatly favored small-scale fractionated research with a relatively brief time span.

It is therefore not surprising that psychological research in the Galtonian style did not establish its claim to social relevance on the basis of sophisticated long-term statistical studies. There was a highly effective shortcut available to it – to hitch its wagon to prevailing preconceptions regarding causation in human affairs. Those preconceptions prescribed that human interaction was to be interpreted as an effect of stable, inherent causal factors characterizing each separate individual. The most important of these factors were defined as psychological in nature, intelligence, and temperament to begin with. Such factors were causal in the sense that they set rigid limits for individual action and were themselves unalterable. The link to hereditarian dogma was quite direct, and many of the major proponents of the Galtonian style of research in psychology were also keen eugenicists.[37] The statistical patterns they constructed were interpreted in terms of hereditary individual traits. For example, a normal distribution would be sought after because of a prior belief that this was the characteristic distribution of fixed biological traits. Psychological knowledge claims in the Galtonian mold generally derived their significance from their reference to inherent properties that were built into each individual, although they could only be assessed by assuming them to be of essentially the same nature in all individuals. The structure of such claims remained unaltered where the original geneticist basis was dropped and replaced by a noncommittal reference to "traits."

The transformation of the old psychology of group differences by Galtonian perspectives led to an important development in the nature of the groups that were the carriers of psychological attributes. Originally, psychologists had simply adopted the categories that defined group membership from everyday life, age and sex being the prototypical examples. But the Galtonian use of statistics greatly facilitated the artificial creation of new groups whose defining characteristic was based on performance on some psychological instrument, most commonly an intelligence test. A score on a mental test conferred membership in an abstract collectivity created for the purposes of psychological research. This opened up untold vistas for such research because psychologists could create these kinds of collectivity ad infinitum and then explore the statistical relationships between them.

Origins of the treatment group

In chapter 5 we noted that in the first half of the twentieth century the investigative practice of American psychology came to rely increasingly on the construction of "collective subjects" for the generation of its knowledge products. Such constructed collectivities were not to be found outside psychological practice but depended for their existence on the intervention of psychologist-investigators. Three types of these artificial collectivities were distinguished: those that were the result of averaging the performance of individuals subjected to similar experimental conditions, those that were constituted from scores obtained by means of some psychometric testing instrument, and those that were produced by subjecting groups of individuals to different treatment conditions.

Whereas the first type of aggregate represented an outgrowth of traditional experimental procedures, the second type clearly owed its existence to the demands placed on psychological investigative practice by its one major external market, the education industry. However, this market also appears to have been responsible for the emergence of the third type of constructed collectivity, the treatment group.

We have seen that what the educational efficiency experts expected from psychological research was, first, methods of measurement that would permit a comparison of results, and second, a rationalized procedure for assessing the effects of various kinds of administrative intervention. Mental tests satisfactorily fulfilled the first expectation, but what about the second? In order to assess the relative efficiency of alternative educational programs, a way of comparing groups exposed to these programs was needed. Yet traditional experimental psychology, because of its focus on the individual, was of little help in this matter.[38] Educational researchers therefore began to find their own ways of coping with the problem.[39]

The pioneer was W. H. Winch, an English school inspector, who was interested in comparing the relative effectiveness of various classroom conditions on such matters as mental fatigue and transfer of training.[40] To do this, he subjected equivalent groups of children to different conditions, taking the relevant measurements before and after the intervention.[41] His example was quickly followed by others investigating such topics as the relative merits of different methods of instruction or the effects of speed on learning.[42]

There was nothing particularly surprising about this development. This was a period during which there was a pervasive tendency to see school problems as problems of work efficiency, a point to which we shall return. Schools were under pressure to work as efficiently as possible. Because school work was carried out in groups (i.e., school classes), the assessment of its efficiency depended on aggregate measures. In this framework the work of the individual child was of interest only insofar as it contributed to or detracted from the overall effectiveness of the work unit.

What is interesting about these early classroom studies is that they took place in an institutional environment that allowed for a certain systematic manipulation of conditions. Thus, studies of work efficiency readily took on an experimental form. The possibility of investigating the effect that variations in the conditions of class work had on measures of output provided a strong practical incentive for combining the use of group data with the experimental method. It was not long before the first examples of studies using the treatment group approach began to appear in journals devoted to basic research.[43] Although such studies were very similar, both in form and content, to the studies conducted in schools, their authors attributed a much broader significance to their results. It was not just the effect of specific classroom conditions on work at school that was now supposedly studied but "fatigue" in general and "learning" in the abstract.

The subsequent history of treatment group methodology ran a rather different course in "applied" psychological research, conducted in institutional settings, and in laboratory research. In the former, there was an initial period of enthusiasm. Within a decade of Winch's first publication the procedure was being sold to American school superintendents as the "control experiment" and touted as a key element in comparing the "efficiency" of various administrative measures.[44] Five years later, W. A. McCall, a longtime associate of Thorndike at Teachers College, Columbia, gave the method its textbook codification for educational research.[45] McCall's justification of experimentation in schools was strictly financial: "Experimentation pays in terms of cash." In one illustrative case he estimated "the value of the increased abilities secured" at ten thousand dollars.[46] Technically, McCall's text was quite sophisticated for its age, expounding on randomization and complex experimental designs two years before the publication of R. A. Fisher's much better-known text.[47]

In spite of these promising beginnings, the history of treatment-group methodology in research in education and educational psychology was not exactly a march of triumph. In the *Journal of Educational Psychology* the use of groups constructed along these lines was confined to 14 to 18 percent of empirical articles during all three periods sampled for the present study, 1914–1916, 1924–1926, and 1934–1936. During the later periods experimental studies in the field of education would be more likely to be published in new journals, like the *Journal of Experimental Education*. In due course, however, a certain pessimism about the prospects of experimental research in education began to set in. The claims made on behalf of the quantitative and experimental method had undoubtedly been wildly unrealistic, and in the light of changing priorities the illusions of the early years were unable to survive.[48]

In the *Journal of Applied Psychology* the use of treatment groups was even less impressive, advancing from 1 to 6 percent of empirical studies between 1917–1920 and 1934–1936. Clearly, there was almost no demand for this type of research practice outside the educational system, and in the latter the demand was limited. This outcome suggests the operation of institutional constraints that tended to impose a limitation on experimental research in institutional settings.[49] Such research had to be carried out with the approval of authorities who presided over power structures that were for the most part highly bureaucratized. The shape of such research was always at the mercy of institutional priorities and shifts in these priorities. In the field of education the introduction of comparative experimentation coincided with the heyday of the "cult of efficiency," but once the influence of this movement waned the impetus for the continued expansion of that kind of research was gone.

More generally, there are always limitations on the utilization of an experimental approach in the context of authoritatively structured institutions geared to practical goals and interests. Not only can experimentation constitute a practical nuisance, but more important, it easily becomes seen as a threat to vested interests and the traditional practices of established power groups. Although psychologists could help both with the selection of individuals for preestablished programs and with the selection of the programs themselves, the former function was far safer in that it lacked any potential for competition with the power of established authority to dispose over institutional conditions.

The situation was quite different in the area of laboratory research where experimentation could safely be used as the preferred method. In the safety of university laboratories a kind of experimentation that had originated in relatively mundane practical concerns developed into a vehicle for maintaining fantasies of an omnipotent science of human control. In their original school settings experiments based on treatment groups could be of some practical value because the institutional context provided a known and consistent framework for applying the results. But the more ambitious American psychologists were not content with the humble role of psychological technician. To them the performance of children on school tasks under different conditions was simply one instance of the operation of generalized "laws of learning" that manifested themselves in the entire gamut of human behavior. This faith enabled them to continue to use a style of experimentation that had already proved its practical usefulness and to reinterpret the results of such experimentation so they now provided evidence bearing on the nature of the fundamental laws of behavior.[50] They imagined that what they observed in these experiments were not specific effects depending on a host of special conditions but the operation of a few highly abstract principles of organismic learning or forgetting.

The relationship of such a science to practical efforts at optimizing human performance was seen very much like the relationship between physics and engineering.[51] Any practical success at increasing the efficiency of school learning, for instance, was thought to depend on the operation of general laws of learning that existed independently of any context. The abstract laws that psychology sought were conceived to be analogous to physical laws. Physics, and especially astronomy, had had to confront the clash between faith in the generality of physical laws and the undeniable variability of observations. That problem was solved to everyone's satisfaction by showing that errors of observation obeyed statistical regularities. In the case of abstract psychological laws there was the additional problem that these laws would be instantiated in the variable behavior of individuals, human or otherwise. It was widely assumed that that source of variability could be reduced to statistical order by an analogous application of the calculus of error. This assumption allowed psychologists to retain their faith in the existence of abstract, universally valid, laws of psychology in the face of daily confrontations with human and animal variability in investigative situations.

Unquestioned faith in the existence of abstract underlying laws produced an interpretation of inconsistent observations in terms of an assumed continuous variability of individual behavior on a set of common dimensions. This interpretation, however, blurred the distinction between generalizations that expressed a truly general relationship and generalizations that were based on no more than a statistical construction superimposed on heterogeneous material. If this distinction could be avoided – and it certainly was avoided[52] – one could turn the variability of one's material into an advantage. If one worked with groups of subjects, one could always structure the outcome in terms of some overall central tendency and some measure of individual deviation from that tendency. This possibility made it terribly easy to produce generalizations that could be held to express psychological "laws," provided one operated in a scientific culture that made no distinction between statistical and psychological reality. The rising popularity of group experiments among practitioners of "basic" psychological research was no doubt promoted by this feature.

The basic science that was sought was really an abstract form of the practice of controlling the performance of large numbers of individuals. Its fundamental laws were to take the form of relationships between environmental interventions and changes in the average response to such interventions. For such a science the use of treatment groups in practical contexts would have an exemplary quality. Transferring their use to the laboratory allows one to construct a model of the kind of world that the laws of this new science presuppose. It is a world in which individuals are stripped of their identity and their historical existence to become vehicles for the operation of totally abstract laws of behavior.

The treatment group was uniquely appropriate for establishing the knowledge claims of a science constructed out of relationships between abstract external influences and equally abstract organisms. Taking the same group of subjects through the same set of environmental variations could also be used for this purpose, and indeed it was. But this approach provided subjects with a bit of history, even though it was only a specially constructed experimental history. This meant that the results could not be unambiguously interpreted in terms of environmental effects on ahistorical subjects.[53] Only the treatment group could provide a practical construction that was a pure reflection of the kind of theoretical construction that the so-called science of behavior favored. As the use of this foundation for knowledge claims became more widespread, the empirical basis for questioning the prevailing shape of psychological theory became narrower. In terms of their overall structure experiments increasingly resembled a practical enactment of what theory presupposed.

8

Investigative practice as a professional project

Interpreting the professional project

The previous chapters have provided abundant evidence for the fact that the historical development of psychological research practice did not proceed along a single track. At the very least, there were two different lines of development. By the 1920s the Wundtian style of experimentation, with its roots deep in the philosophical and scientific traditions of the nineteenth-century German university, seemed to constitute a deteriorating research program within American psychology. As it was fast losing its appeal for all but a few practitioners, the Galtonian research program, strongly linked to practical rather than academic concerns from the beginning, was moving from strength to strength. Not only was it extending its appeal with every passing year, but, as we have seen, it proved itself capable of generating exciting methodological innovations that promised to extend the scope of scientific psychology far beyond what had hitherto been thought possible.

We have also seen that the galloping success of the Galtonian program was in no small measure due to its very direct link to the demands of a significant extradisciplinary market for its products. Nevertheless, it would be a mistake to see the development of a certain style of investigative practice as an essentially passive adaptation to external social requirements. That might have been true of certain individuals, but it was not true of the response of the discipline as a whole. Even on an individual level most practitioners recognized that practical successes would be short lived if they threatened to undermine the prestige of the disciplinary enterprise they represented. If the practice of mental testing, for example, proved to be transparently unreliable, it would do the practitioners no good in the long run, no matter how spectacular the immediate yield might appear to be. The practical effectiveness of psychologists depended quite crucially on their ability to maintain and enhance whatever standing as experts they had managed to achieve. That standing, in turn, depended almost entirely on their ability to convince their publics that they represented the sacred spirit of Science. Legitimating their practices in terms of supposedly universal and eternal principles of scientific method was therefore a permanent preoccupation – often to the point of obsession – for all farsighted practitioners.

What exactly was the nature of the product that American psychologists were beginning to market successfully from about 1910? Were their mental-tests goods like material commodities that lost their intrinsic link to the producer once they were sold? Obviously not, for attempts to treat tests in this way led to justified charges of their "misuse" at the hands of "untrained" personnel. Clearly, what the psychologists were marketing were their own skills and their own special competence of which the test materials were merely the external mark. The product that their society valued was *professional expertise*. However, this was a pretty abstract category, and any specific instance of it still had to be established as being the genuine article and not some sort of counterfeit quackery. In other words, the social acceptance of their expertise required that psychologists legitimate their claims in terms of certain widely accepted criteria, which were based on prevailing conceptions of scientific method. It was not enough that their methods yield results that were useful to socially dominant groups; the methods themselves had to be seen to be rational, which, in that particular historical context, meant that they had to bear the hallmarks of science.

In developing their investigative practices psychologists were not only producing instruments for the solution of technical problems, they were also establishing and improving their own status as professionals and as scientists. In a society that places such high value on the methodology of science, the practice of anything recognized as belonging in that category will reflect positively on the prestige and social status of the practitioner. Investigative practices are neither neutral nor disembodied. They carry a social value, and they are inseparable from the social standing of those who enact them.

By the early twentieth century the use of certain investigative practices reflected not only on this or that individual but on a group with the social identity of "psychologists." What the individual investigator was seen to be doing had potential social consequences for others who shared his or her professional identity. Practitioners within the discipline were increasingly tied to each other by their participation in a *professional project* whose fortunes would potentially affect all of them.[1] This project involved matters of social status, of collective social mobility, of competition for scarce resources, of boundary negotiation with neighboring disciplines, of demarcation from quacks and lay practitioners, as well as related concerns. The kinds of investigative practices with which psychologists were associated were inseparable from their professional project because these practices played a role in virtually every aspect of the project. They served to distinguish psychologists from certain neighboring disciplines and professions with which there was actual or potential competition, and they served to draw a sharp line between experts and laymen.[2] Above all, however, investigative practices were strongly implicated in that aspect of the professional project concerned with legitimating the scientific credentials of the discipline.

The criteria for being recognized as scientific had relatively little to do with how the established sciences actually achieved their successes. That involved a set of very complex issues, which remain controversial to this day. What was socially important, however, was the widespread acceptance of a set of firm convictions about the nature of science. To be socially effective it was not necessary that these convictions actually reflected the essence of successful scientific practice. In fact, the most popular beliefs in this area were based on external and unanalyzed features of certain practices in the most prestigious parts of science. Such beliefs belonged to the rhetoric of science rather than to its substance.[3] They generally involved the elevation of certain unexamined and decontextualized features of scientific procedure to the status of fetishes. Such sacred and unquestioned emblems of scientific status included features like quantification, experimentation, and the search for universal (i.e., ahistorical) laws. A discipline that demonstrated its devotion to such emblems could at least establish a serious claim to be counted among the august ranks of the sciences.

The contribution of investigative practices to the professional project of psychology involved two sets of problems with often diverging implications. On the one hand, there was the need to develop practices whose products would answer to the immediate needs of socially important markets. But on the other hand, there was the need to establish, maintain, and strengthen the claim that what psychologists practiced was indeed to be counted as science. These two requirements could not always be easily reconciled, and so it was inevitable that there was conflict within the discipline with some of its members placing relatively more emphasis on one or another of these directions.[4] But in the long run the two factions depended on one another, rather like two bickering partners in a basically satisfactory marriage.[5] For the very term "applied psychology" reflected the myth that what psychologists put to use outside universities was based on a genuine science, much as engineers based themselves on physics.[6] Without this myth the claims of practitioners to scientific expertise must collapse. But at the same time, the pure scientist in the laboratory could often command much better support for his or her work if it was seen as linked to practical applications by a society for which utilitarian considerations were paramount.[7]

For American psychology the period between the two world wars was marked by considerable dissension between its "basic" and "applied" wings, but also by determined efforts to find a compromise that would satisfy both parties. These conflicts were quite clearly reflected on the level of research practice. One orientation, represented by a few prominent individuals and by prestigious publication outlets like the *American Journal of Psychology* and the *Journal of Experimental Psychology,* strongly favored experimental practices and shunned the use of mental tests as well as natural situations and populations. The other orientation, strongly represented in research publications like the *Journal of Applied Psychology* and the *Journal of Educational Psychology,* favored mental tests and made only limited use of experimentation.

The two orientations seemed to draw selectively on different criteria for establishing the worth of their investigative practices. Experimentalists were able to draw on the apparent analogy between what they were doing and what physical scientists were doing.[8] There was no question that engaging in experimentation for the purpose of discovering universally valid natural laws constituted scientific practice. On this kind of criterion much of what the applied psychologists were doing seemed to be of doubtful respectability but the latter group could always point to the relevance of their work for immediate social and institutional problems.

Nevertheless, this obvious divergence should not blind us to a counteracting convergence that was operating at the same time. One important expression of this convergence is provided by the experimentalists' gradual switch from single-subject research to the statistically constituted group data that had always been favored in the more "applied" kind of research. This switch was alluded to in chapter 5. Once experimentalists had accepted the need to organize their work around the production of group data, the gap that separated the investigative practices of the two disciplinary subgroups had been significantly narrowed and a basis for a new disciplinary consensus had been established. In the next two sections we will look more closely at the dialectical interplay of converging and diverging tendencies in the development of investigative practices during the crucial interwar period.

Divergence of research styles

In chapter 5 we distinguished between two kinds of collectivities that were employed in the generation of group data. "Natural" groups were defined by culturally given categories, whereas "artificial" groups owed their definition to the investigator's intervention. Examples of the former were sex and age groups, while the latter were exemplified by groups defined in terms of scores on some psychometric test or by the experimental treatment they had been given. Two general tendencies were noted (Tables 5.3 and 5.4): an overall trend, during the interwar period, for group data of all types to replace individual data; and a much greater tendency to use group data in the applied literature than in journals of "basic" research.

If editors and contributors to these two sets of journals were selectively emphasizing different aspects of psychology's professional project, as discussed in the previous section, we would not expect to find the same mix of group data in both sets. Data pertaining to "natural" groups have a manifest social relevance, which is due to their display of knowledge in terms of culturally familiar categories. For example, one could always count on a significant lay interest in information that pertained to differences among children varying by age, background, and sex. The production of such information would be high on the agenda of those who emphasized the role of psychology as a socially useful discipline, and we would expect the "applied" journals to function as their outlet.

Data generated by the use of experimentally constituted groups, however, did not generally exhibit the same degree of manifest social relevance. Psychologists certainly claimed social relevance for some of the statistical data generated in the course of experimentation, but such claims tended to be more controversial and less accessible to lay understanding than knowledge claims based on the more descriptive use of "natural" groups. What the production of experimental data did have in its favor was the analogy with the activity of "real" scientists.

The differences between "basic" and "applied" journals in respect to their selective emphasis on different types of group data are summarized in Table 8.1. Two points should be noted prior to decoding the significance of these figures. First, the percentages entered in the cells of Table 8.1 refer to the ratio of studies based on each kind of group to the total number of empirical studies published in a particular set of journals. Thus, the sum of the percentages in each row is always less than 100, the difference being made up by studies that used no group data or a mixture of group and individual data. Obviously, the sum of the percentages in a row will be higher the greater the proportion of studies that relied purely on group data. However, the patterns in the overall use of group data have already been discussed in conjunction with Table 5.3 and 5.4. What is of interest now are the differences in the relative use of different kinds of group data.

Second, it should be noted that our coding category of "psychometric groups" does not provide an index of the simple use of mental tests for purposes of social selection. By the 1920s it is actually the "natural groups" category that is mostly made up of studies of this type. What the "psychometric groups" category represents are correlational studies in which distributions of scores on different tests or scales are compared with each other – that is, studies in which the categories of research are entirely generated by the research activity itself. Such studies were of primary interest to psychologists themselves and were at one remove from the apparently immediate social relevance of natural-group studies.

Table 8.1. *Use of different types of group data*

Type of group	Journal of Educational Psychology	Journal of Applied Psychology	Journal of Experimental Psychology	American Journal of Psychology
1914–1920				
Natural	25	37	6	0
Psychometric	19	24	9	3
Experimental	31	15	23	23
1924–1926				
Natural	38	35	16	11
Psychometric	29	36	3	5
Experimental	25	11	25	19
1934–1936				
Natural	45	34	12	8
Psychometric	22	39	2	4
Experimental	27	20	48	44

Notes: Number of group data studies based on each type of group as a percentage of the total number of empirical studies for each time period. In the case of a minority of studies that reported results based on more than one type of group, a coding priority rule was used that ordered the categories of groups in reverse order from the order in which they appear in the table. Thus, if any part of the study relied on experimental groups, that would be the code assigned, irrespective of other types of group that might also occur in the same study. If a psychometric group was used, that category took precedence over natural group. See appendix for explanation of the 1914–1920 time period.

It is clear that there are some striking differences between the "applied" and the "basic" research literature. In the first place, the two applied journals, the *Journal of Applied Psychology* and the *Journal of Educational Psychology,* are much more likely to publish studies with results that are directly attributed to "natural" (i.e., real-life) groups. These journals clearly maintain the tradition of a socially relevant science that attempts to provide knowledge pertaining directly to categories of persons defined by culturally accepted criteria. In the case of one of these applied journals, the *Journal of Educational Psychology,* the exclusive reliance on data pertaining to natural groups even increases from a quarter to nearly a half of all the published empirical studies in the course of the period under review.[9] The attribution of findings to existing social categories is clearly very marked in the applied literature but relatively weak in the journals devoted to basic experimental research.

However, this does *not* mean that the applied literature lags behind the basic literature in regard to the use of artificial groups. In the *Journal of Educational Psychology,* one-half of all the empirical studies involve attribution of findings to artificial groups from the earliest time period onward. The *Journal of Applied Psychology* reaches this level of reliance on artificial groups in the 1920s, but the basic research journals do not reach it until the following decade. By that time the *Journal of Applied Psychology* has reached a level (59 percent) that suggests a continuation of the trend toward even greater reliance on artificial groups in the future. There appears to be an overall trend in three of the four journals for the use of artificial groups to be distinctly more prevalent at the end of the twenty-year period than it was at the beginning. Interestingly, the exception is the *Journal of Educational Psychology,* which starts with the highest incidence of artificial groups but then moves in the direction of increasing the use of natural groups.

These data make it possible to give a more differentiated account of the increasing tendency to use group data that was noted in chapter 5. In the main, this now emerges as a tendency to make more use of artificial groups in the attribution of data. Increasingly, psychological research constitutes its own collective subjects, which are defined by criteria that are different from the criteria used to group individuals in everyday life. The discipline is busily constructing its own set of categories, which its research practice then imposes on the individuals studied. More and more of psychological knowledge comes to refer to a world in which people are grouped by specially constructed categories that may have only a vague and tenuous connection with the categories that are used to distinguish people in ordinary life.

We can regard any set of psychological research data as a product of two sets of knowledge interests. On the one hand, there is an interest in producing the kind of knowledge that appears to be practically useful to certain potential consumers of the knowledge outside the discipline. But on the other hand, there are interests that aim at the advancement of the discipline, both in a cognitive and a social sense. What is desired here is knowledge that primarily will appear to further the cognitive and technical control of the discipline over its subject matter and will improve its status among the sciences.[10]

These two sets of interests are not necessarily antagonistic, although they may at times become so. Intradisciplinary interests may in fact be a reflection of broader societal interests, and may simply be projecting a much longer time frame for the achievement of broadly the same goals. But at any one time the two sets of interests are likely to be clearly distinguishable and to entail different *immediate* consequences for research practice. Thus, the mixture of investigative practices to be found in the discipline, or one of its subdisciplines, at a certain time can be regarded as the product of a particular balance between these sets of interests.

In the field of educational psychology the pressures to produce knowledge of immediate relevance to real groups of children appear to have been highly effective. The relatively close integration of psychological practice with educational institutions meant that the target populations for research were increasingly defined by those institutions. This integration did not occur at the expense of psychologically defined collectivities, however, but at the expense of the last vestiges of traditional psychological research in this area. Educational psychology accommodated its practices quite fundamentally to the requirements of its market, but in doing so it actually advanced the disciplinary project by making psychological categories a factor to be reckoned with in some important real-world institutional contexts.

By contrast, psychologists in the industrial field had to face quick disillusionment, when the expected profitability of mental tests was not immediately apparent.[11] They also faced a major threat from untrained practitioners.[12] Although attempts to demonstrate the direct relevance of findings to real groups continued unabated, the investigative practice of applied psychology during this period seems to have been marked by redoubled efforts at improving its technical sophistication. This would not only increase the distance between professionals and nonprofessionals but might also improve the predictive value of instruments on which any future practical success of the area depended.

But these were special subdisciplinary problems. Returning to the overall pattern of disciplinary practices, we still have to take into account certain differences in the use of artificially constituted collectivities for purposes of data attribution. The major distinction here is between psychometrically and experimentally defined groups. In the former, the group of subjects to whom the research findings are attributed is defined by its possession of certain psychological qualities *elicited* by the psychologist's intervention. By contrast, experimental groups are defined by their having undergone a certain pattern of experimental intervention. The difference between the psychometric and the experimental type depends on the nature of the psychologist's intervention, which depends in turn on the way in which that intervention is interpreted by the investigators. In the one case the observed activity of the subjects is considered essentially as an *effect* of the intervention; in the other case it is considered essentially as an *expression* of some quality of the subjects.

Table 8.1 indicates a massive difference between the applied and the basic research journals in the use of investigative groups constituted by psychometric intervention. Such groups occur much more frequently in the former than in the latter journals, and, if anything, this divergence appears to increase somewhat in the course of time. Conversely, there is more use of experimental intervention as the defining characteristic of investigative groups in the basic-research journals. Our choice of journals appears to have provided information that suggests that there were marked differences within the discipline in regard to the uses of group data. For some, such data were useful in the overall assessment of experimental intervention because it enabled one to iron out individual differences. But for others such data were needed precisely because it was desired to have some basis for placing individuals in relation to the group.

If we distinguish between the research published in the applied journals and that published in the basic journals, it is clear that during the interwar period it was only the latter group that was undergoing something like a revolutionary development. The pattern for applied-research styles had been essentially established by the end of World War I, and in the ensuing years no fundamental changes occurred. This was basically a Galtonian style of research concentrating on the distribution of psychological characteristics in natural or psychometrically constituted populations. Thus, by this time there were two quite divergent styles of psychological research in existence. One worked with data from individual subjects reacting under laboratory conditions, the other with populations surveyed statistically. The one had the weight of tradition and the mystique of the laboratory behind it, the other was buoyed up by apparent practical success and immediate social relevance.

Emergence of a compromise

If the professional project of American psychology contained contradictory elements, this was entirely due to the fact that it had to come to terms with inconsistent social demands. The material well-being of the discipline depended *both* on its ability to deliver results of immediate practical utility *and* on the successful cultivation of its scientific image. A tendency toward a division of labor resulted, with one part of the discipline emphasizing direct social relevance and the other, scientific respectability. But it would be a mistake to see only the divergence on the surface and to ignore the more subtle unifying tendencies that were also emerging.[13] In fact, the two major directions were not totally distinct but showed an increasing degree of interpenetration. Faced with divergent social demands, it was only to be expected that practitioners would become more and more adept at finding solutions that would reconcile these demands.

Thus, the new generation of educational psychologists coming to prominence in the early years of the twentieth century was able to eclipse the previous efforts of G. Stanley Hall at "child study" partly because they commanded a more sophisticated scientific rhetoric. It was not just practical relevance that guaranteed success, but practical relevance covered by the mantle of science. In this case the science in question was grounded in the statistical biometry of the Galton–Pearson school. It is true that the psychologists who developed mass mental testing on this basis were responding to administrative requirements, but their successful use of the rhetoric of science enabled them to insert their own categories into the administrative process and so to advance their professional project.

These developments are relatively clear, but during the interwar period there emerged another kind of synthetic practice that affected experimentalists rather than mental testers. In this case it was the impressive practical successes of Galtonianism that began to influence the practice of psychological experimentation.

Laboratory methods were not abandoned in favor of mental testing, but, increasingly, laboratory data were attributed to groups of subjects rather than to individuals. This trend is particularly noticeable at the end of the period. For the two experimental journals the use of group data in a laboratory context doubles between the mid-1920s and the mid-1930s (see Table 8.1). We have previously noted that there was a corresponding decrease in the attribution of experimental results to individuals during this period (chapter 5). In other words, experimentalists were now less inclined than they had been traditionally to make knowledge claims based on the behavior and experience of individuals and more inclined to base these claims on group averages.

This tendency implied a fundamental shift in the function of the psychological experiment. In the traditional approach, going back to Wundt and other nineteenth-century pioneers, the objects of psychological experiments were individual psychological systems and the use of more than one subject constituted a replication of the experiment. In the new approach the experiment was conceived from the beginning as having a group of subjects as its object. This entailed a Galtonian conception of "error" as involving the variability of performance between individuals, whereas in the traditional approach error involved the variability of response by an individual psychological system. The new kind of experiment no longer sought to analyze or characterize the mode of functioning of individual psychological systems directly. Instead, it sought to establish a statistical norm of response from which individual responses were regarded as deviations. In this respect the new kind of experimentation was much closer to mental testing. Where it differed from the latter was in the retention of experimental control and variation of conditions.

The new approach to psychological experimentation was therefore a hybrid of investigative styles that had previously been established in the discipline. For this reason we may refer to it as a neo-Galtonian style.[14] It is undoubtedly experimental, but it employs experimentation in a Galtonian framework.

What factors might have been involved in the development of this new investigative style? As has already been observed, the apparent successes of an applied psychology that was largely based on a Galtonian use of statistics could hardly have failed to make an impression on the more traditional sections of the discipline. Increasingly, the latter accommodated themselves to the goal of making predictions about the average response of aggregates. It was this kind of knowledge that seemed to lead to a socially relevant psychological science.

The 1920s were a period during which the older ideals of an academic discipline devoted to the painstaking examination of individual mental processes crumbled before the new goal of putting psychological prediction in the service of large-scale social control. By 1921 three-quarters of the psychologists appearing in *American Men of Science* listed applied interests. More fundamentally, the academic science of psychology was becoming thoroughly penetrated by practical categories of social thought.[15] This was only too clear to a few traditionalists who wrote about the new developments in the tone of critics left behind by the tide.[16] Some, like E. G. Boring, attempted to stem the tide,[17] but their efforts only served to emphasize the direction in which the discipline was moving.

Among the contextual factors that favored these developments, two are well known. First, psychology was not alone during this period in showing a tendency for practical technology to usurp the name of science. In physical science the practical expert was frequently to be found in a preeminent role and the goals of scientific research were often equated with the goals of engineering.[18] Given the prestige of physical science, analogous developments in psychology must have seemed highly legitimate. Second, we are dealing with a period during which large funds began to be available for psychological research from a new source, the private foundations. In particular, the Laura Spelman Rockefeller Memorial moved into the area of social research on a big scale, and psychologists could benefit from this largesse provided their work seemed relevant to the clearly defined goals of those in charge of the funds. What the latter were interested in was "knowledge which in the hands of competent technicians may be expected in time to result in substantial social control."[19] To many it seemed clear that if experimental psychology was to contribute to this kind of knowledge it would have to drop its old knowledge goals and devote itself to the discovery of universal "laws of behavior" that might be useful in charting projects for social control.[20]

What was attractive to experimentalists about the new methodology was that it allowed them to sidestep the problem of individual differences. In the traditional approach such differences were generally handled in three ways. In the first place, there were persistent attempts to refine the control of conditions so that differences among individuals would be minimized or even eliminated. If this failed, one attempted to find reasons for the peculiarities of individual response patterns – special features of the person's visual apparatus or past experience, for example. Finally, there was always a strong tendency in the traditional approach to limit the scope of experimental psychology to the investigation of those phenomena for which differences among individuals could be successfully minimized or rationally accounted for. This approach meant a very serious restriction on the expansion of the discipline and a refusal to pursue the investigation of the kinds of psychological problems that were relevant to potential lay consumers of the products of psychological research. By subjecting individual differences to statistical treatment as error variance, the new experimental psychology was able to cast aside these restrictions and to make knowledge claims about "laws of behavior" that were supposed to apply across individuals (and also across situations). Aggregate data provided the empirical basis for such "laws," so that they need not be instantiated in the behavior of a single living individual. The locus of operation of abstract psychological laws was an abstract individual who was the product neither of nature nor of society but of statistical construction.

The purpose of psychological experimentation had therefore undergone a change. In the older kind of practice one manipulated experimental conditions in order to test hypotheses about the processes going on in individual psychophysical systems. Now, the direct purpose of experimentation was to make predictions about how certain variations in conditions affected the response of an abstract individual. Because in practice such an individual was statistically and not psychologically real, questions of psychological inference very easily became transformed into questions of statistical inference, a development to which we shall return in the following chapter.

A comparative perspective: The German case

The scope of our historical interpretation has been confined to data derived from a single social context, the American one. But an examination of analogous aspects of the investigative practices of psychology in a different social context during this period is possible because there was a considerable amount of psychological research in Europe as well. However, only in Germany was the sheer volume of work large enough to allow us to perform a meaningful statistical analysis that would be comparable with the analysis we have made of the American research literature. German psychology will therefore provide our comparison case.

As before, the journal literature in different areas of psychology will be examined for its reliance on statistical data based on aggregate data from groups of subjects. The first area to be looked at is that of industrial psychology, the area in which the national differences turn out to be least marked. In sample volumes of the journals *Industrielle Psychotechnik* and *Psychotechnische Zeitschrift* for 1924–1927, the proportion of empirical studies relying predominantly on group data turns out to have been consistently between 60 and 65 percent. Much the same is true (60 percent) for *Industrielle Psychotechnik* in the years 1934–1936. (The other journal ceased to be a significant outlet for empirical studies.) These figures are lower than the corresponding figures for the *Journal of Applied Psychology,* which exceeded 80 percent (Table 5.4). The difference is probably due to the fact that at this time parts of German industry were particularly interested in the rationalization of production through the use of what was to become known as human engineering. This often involved intensive individual studies rather than group data.[21] Nevertheless, the administrative context that provided the bulk of the market for industrial psychology appears to have ensured a preference for statistical group data in both the American and the German case.

When we turn to the educational literature the German–American differences become rather more pronounced. In the *Zeitschrift für pädagogische Psychologie* the proportion of empirical studies based predominantly on group data lies at 53 to 54 percent both for the period 1911–1913 and 1924–1926. (The number of empirical studies is too low for meaningful comparison in the 1934–1936 period.) In the *Zeitschrift für angewandte Psychologie,* whose applied empirical studies fell almost entirely into the field of child and educational psychology, the proportion of such studies relying predominantly on group data was 56 percent for the period 1915– 1916 and 53 percent for the period 1924–1926. These figures are not only considerably lower than the corresponding figures for the *Journal of Educational Psychology* (75 percent for 1910–1912 and 91 percent for 1924– 1926), but they also show no tendency to increase with time.

What kinds of factors might be at work here? One possibility that must be considered is that the appropriate statistical techniques for the analysis of psychological group data were more readily available to American investigators, because the primary source of these techniques was in Great Britain. Such techniques would enhance both the utility and the apparent scientific status of group data, and so might well encourage their use. Now, it is true that by and large the level of statistical sophistication tends to be higher in the American than in the German applied literature of this period. However, this seems to be less a question of the availability of the relevant information than of its reception. The relevant technical information was widely disseminated through such vehicles as well-known textbooks and publications in the specialist journal literature.[22] The influence of Galtonian conceptions of human variability is by no means absent in the German research literature of the period, but it never remotely approaches the overwhelming impact it had on the parallel American literature.

In both Germany and the United States the market for the products of educational psychological research was based on an expanding system of universal education that was adapting itself to the requirements of industrial capitalism. In both cases the effects of this market can be detected, not only on the level of psychological content, but also on the level of investigative practices. These frequently become geared to the production of a kind of psychological knowledge that is relevant to administrative decisions concerning the selection of persons and of programs in terms of the criterion of performance efficiency.

However, the social situation in Germany appears to have imposed far greater limitations on this process than in the American context. In the latter case education had never been an important source of social status as it had been in nineteenth-century Germany, where rigid segmentation among different types of schools was intimately tied to rigid social class distinctions.[23] In Germany powerful and well-established social mechanisms traditionally governed the selection, both of individuals and of programs, within the educational system. Education officials had a vested interest in maintaining these traditional mechanisms and seldom saw any need to replace or even supplement them with the new mechanisms that applied psychology could provide. The situation did not change substantially during the Weimar period, because of the political weakness of the forces of educational reform.[24] Although a limited market for the products of applied psychology did exist, it bore no comparison to the vast education industry that was the powerhouse of the rapid growth of American applied psychology. Instead of functioning as a repository of preindustrial patterns, as it did in Germany, American education quickly adapted itself to provide an almost perfect reflection of the requirements of the new industrial order.[25] As we saw in chapter 7, the chief agents of this process were the new educational administrators who provided applied psychology with its most important and most reliable market.[26]

By contrast, the main consumers of educational psychological research in Germany appear at first to have been classroom teachers. The teachers' associations (*Lehrervereine*) of some of the larger cities played an extremely active role in promoting the dissemination of the products of this research. In a few key instances they established research laboratories and subsidized psychological research from their own funds. The first of these was the Institute for Experimental Pedagogy and Psychology of the Leipzig teachers' association, founded in 1906. This was quickly followed by similar initiatives in Berlin and Munich; other important centers were Breslau and Hamburg.[27] Here there was a direct collaboration of teachers and academics guided by a common interest in classroom teaching and by the goal of persuading the authorities of the need for educational reform. Unlike administrators, classroom teachers were directly concerned with psychological processes in the minds of individual children and therefore had an interest in psychological research conducted on this basis.[28] This is not to say that group data were irrelevant to their practice, but merely to indicate that these interests were tempered by other concerns. Thus one reason for the difference between German and American educational psychology in the overall pattern of investigative practices may have to be sought in the different professional alliances that were feasible and advantageous in these two cases.

However, one must be careful to avoid the kind of oversimplification that would result from considering the educational system as a purely external market for the discipline of psychology. The key figures in the adaptation of psychological investigative practices to the requirements of this market were after all themselves products and functioning parts of the educational system, by virtue of their academic qualifications and appointments. Thus, certain characteristics of the system were not experienced by these individuals as external demands but as internalized values. Among those German academics who took the lead in advancing the cause of applied psychology this often expressed itself in a characteristic duality of goals. The same men who pioneered the advancement of psychological techniques as instruments of social control, and of quantitative methods for providing administratively useful knowledge, often devoted much of their energies to the pursuit of philosophical goals that had been sanctified by the humanistic traditions of German higher education.[29] Such split loyalties contrast rather sharply with the single-minded devotion to the ideal of calculated efficiency and rationalized performance that was so common among their American counterparts.

We are therefore led back from a consideration of market conditions, external to the discipline, to the question of intellectual interests operating within the discipline. These become particularly important when we enlarge the scope of our inquiry to consider the influence of the investigative practices of applied psychology on the discipline as a whole. It was shown earlier that in American psychology the fundamental shift from the analysis of psychological processes in individuals to the analysis of the distribution of psychological characteristics in populations had significantly affected investigative practices in the area of "pure" research by the 1930s. Can an analogous effect be demonstrated in the German literature of the time? The answer is yes – in a very limited way. During the years 1934–1935 the proportion of empirical studies relying entirely on group data reached a mere 13 percent in *Psychologische Forschung,* 20 percent in the *Zeitschrift für Psychologie,* and 23 percent in the *Archiv für die gesamte Psychologie.* Although these figures are far below the corresponding American figures, where this approach already accounted for more than half the published studies, they do represent a small but consistent increase over the German figures a decade earlier. For the period 1924–1926 the proportion of studies relying solely on group data varies from 0 in *Psychologische Forschung* to 13 percent in the *Archiv für die gesamte Psychologie.* This approach to psychological research was always less heavily represented in the German than in the American research literature. However, a slight tendency to follow the lead of applied psychology in adopting this approach does eventually become noticeable.

The difference in the historical development of investigative practices in the American and the German case is not an absolute one. Analogous forces seem to be at work, but their relative weight is very different. In both cases the research literature shows a mixture of fundamentally different research practices, which are geared to the production of different kinds of psychological knowledge. In what I have previously called the Wundtian model, psychological research aims to analyze processes located in individuals, whereas in the Galtonian model the concern is with the distribution of psychological characteristics in populations, and the placing of individuals with respect to such group characteristics. Both in American and in German psychology the Galtonian approach is strongly favored by the emergence of an important market for the products of psychological research. Also, in both cases there is a detectable shift in the general research literature toward the Galtonian model.

However, these effects appear to be much stronger in the American than in the German case. Two kinds of factors are likely to account for this difference. In the first place, certain differences in the social role and structure of the educational system had the effect of turning American education into a far more powerful market for psychological products and services than was the case for German education, at least during the first half of the century. But this does not account for the curiously ambivalent response on the part of German academics interested in applied psychology, nor does it account for the extremely weak effect that the Galtonian model of psychological research had on the discipline as a whole. For an explanation of these phenomena we have to consider the way in which members of the discipline defined their professional project. Among American psychologists, there was an early recognition of the fact that the progress of their discipline depended on its ability to supply knowledge that would be useful for the practical management of human affairs. Such early pioneers as G. Stanley Hall and James McKeen Cattell already grasped that the pursuit of this goal would require a switch to research practices that yielded statistical information about defined groups of individuals. Thus, the response to the requirements of the educational market was rapid and unambiguous. This did not mean that there was unanimity within the discipline, but simply that there existed a convergence of external and internal factors that proved decisive for the overall development of the discipline in the long run.

By contrast, German academic psychologists found themselves in a situation where their enterprise was obliged to legitimate itself, in terms of the values represented by a well-entrenched traditional intellectual elite, if it was to prosper.[30] On the level of research practices there was no chance of accomplishing this by means of methods that were seen as a kind of superficial census taking that negated human individuality as well as the traditionally recognized mandate of psychology to explore the individual mind and personality.[31] The weight of such objections would have been all the greater, as the institutional separation of psychology and philosophy had not become the norm at German universities. Academic psychologists often held appointments in philosophy, and in many cases their own intellectual commitments matched these institutional arrangements. Whether their philosophical position owed more to the humanistic tradition or to the newer irrationalist currents did not matter; in either case their distaste for a numerizing approach to human individuality was predictable.

As in the United States, there were pulls on the emerging discipline from a very different direction, that of psychotechnology and its various areas of application. Here, philosophical scruples counted for nothing, and the only tribunal to which statistical techniques had to answer was one guided by norms of pragmatic effectiveness. The question now became one of which set of interests would set its stamp on the discipline as a whole. Both in Germany and in the United States academic psychologists played the crucial role in determining norms of good disciplinary practice. The difference lay in the ties, institutional and economic, that academic psychologists had with potential centers of influence.[32] In Germany the separation between "pure" science and technology, although not as strong as it had been in the nineteenth century, still went very deep, both institutionally and subjectively. Technical studies were segregated in the *Technische Hochschulen,* or Technical Universities, at some of which applied psychology did find a home. However, during the Weimar period at least, the intellectual leaders of the discipline were not to be found there but at the regular universities, where they were immune to any influences from the dirty world of industry. The barriers, both institutional and subjective, against any adoption of ideas or practices from that quarter were formidable. A very different situation existed in the United States, where institutional pressures from a number of sources – university governments; research foundations, both public and private; bureaucracies – all converged to support the orientation of investigative practice toward managerial concerns.

Although it is convenient, for the sake of conceptual clarity, to make analytic distinctions between factors external and factors internal to the discipline, the limitations of such distinctions must be clearly recognized. Internal definitions of disciplinary goals are themselves adaptations to external conditions, and conversely, disciplinary projects contribute to the formation of external markets for their products. From the viewpoint of the process of professionalization, cognitive schemata and technical practices can be treated as a resource,[33] whereas in the present analysis it is precisely these "resources" that are regarded as problematic. The constellation of intellectual interests that constitutes the professional project faces both outward and inward. These interests mediate between the possibilities and demands that confront the community of practitioners and the range of investigative practices accepted as legitimate within the community. These practices in turn circumscribe the kind of knowledge that the discipline is able to produce and define the limits of that knowledge.

9

From quantification to methodolatry

Twentieth-century psychology's infatuation with numbers would not have blossomed into an all-consuming love affair had it not been for the usefulness of statistical constructions based on group data. Statistical constructions represent work done on some raw material that already has a numerical form. However, that form involves more than the counting of heads. The individuals who are counted must also be endowed with countable attributes. Where the attributes are physical, like height or weight, their transformation into numerical form depends on social practices developed long ago, but where one has to deal with psychological attributes the corresponding measuring practices represent a relatively recent development. This development was initially independent of the special use of statistical information that has occupied our attention so far. It is therefore necessary to step backward in time to consider the nature of the practices that made it possible to impose a numerical form on psychological attributes.

What is the difference between one and two?

Because it is rare for human experience and activity to take on a quasi-numerical form spontaneously, psychologists who insist on working with such forms must impose them on their subject matter. One way of doing this is to apply the numerical net post hoc, after the psychological event has taken place and has produced some record. The number of word associations with a particular thematic content can be counted, for instance, either for a particular individual or across individuals. This kind of quantification was attempted by Galton and practiced on a large scale by G. Stanley Hall.[1] Here the production of numerical data depended on the researcher establishing certain thematic categories and then performing a number of equivalence judgments that resulted in unambiguous decisions about the inclusion or exclusion of each judged item for the categories employed. The point is that the numerizing operations were clearly performed by the researcher and not by the human source of the data. The latter's experience and activity was not directly affected by these operations, except in the special case where the researcher was his own data source.

This approach, however, was of limited significance for the subsequent development of quantifying practices in psychology. Much more important was a second approach that attempted to impose a numerical structure on the data of psychology right at their source. The various techniques enlisted in the service of this aim can be regarded as constituting a hierarchy of increasingly complete control of individual behavior. First, a numerically graded stimulus series can be prepared – for instance, a series of weights of increasing heaviness – and the subject's responses to each member, or each pair of members, of the series can be recorded. This was done very early in G. T. Fechner's psychophysical methods,[2] which formed a mainstay of the work of the first psychological laboratories.

However, it should be noted that for a quantitative psychology something more is needed than a numerically graded stimulus series. It is also necessary to limit the subject's responses to these stimuli. A qualitative description of sensations is not what is required here. The experimental subject must limit himself to simple judgments, and his answers must be cast into an unambiguous, digitalized form of the yes–no type. Only under these conditions is it possible to obtain a quantitative function that links the stimulus series to the response series in the kind of relationship that Fechner called Weber's Law. For instance, the subject must agree to limit his responses to alternatives like "heavier versus lighter" or "equal versus unequal" if the goal of this sort of investigation is to be achieved. Because Fechner experimented on himself, the goals of the investigator and the duties of the experimental subject were brought together in a kind of personal union, but in a social functional sense they are distinct. There is an indispensable element of social control wherever the form of subject responses is strictly subordinated to the goals of the investigation, although the social aspect may well be hidden by the voluntary nature of the agreement and by the impersonal nature of the requirement.

Sometimes the practice of controlling the form of the subject's response in the interests of quantification did lead to debates among experimentalists in which the social basis of the practice emerges clearly enough. Numerization required that only responses of the dichotomous yes–no type be given legitimacy. But what was one to do about human individuals' spontaneous and persistent tendencies to break out of this straightjacket, if only by giving responses in the "doubtful" category? There were two alternatives. One could prevent the subject from taking such liberties by insisting that all answers be either yes or no (the method of right and wrong cases), or one could be more permissive and allow doubtful responses, but afterward manipulate the data so that the effect of such responses was washed out.[3] The simplest way of doing this was Fechner's method of dividing the "doubtful" items among the legitimate (i.e., yes or no) categories. Whatever technical sophistication was added in the latter case, there had to be a basic decision on how far the investigator ought to push his authority in the experimental situation. Such an issue could only arise within an existing social framework. The production of quantitative data involved social practices of investigation at a very fundamental level.

More generally, social numerizing practices of the early experimentalists linked theory and data in a tight methodological circle. For Fechner, and those who followed him, the confinement imposed on the subjects' responses seemed appropriate and justified because "sensations," which was what they considered themselves to be studying, were assumed to constitute psychological atoms or elements that could be incrementally added if they were of the same kind.[4] The instructions and response categories imposed on individuals acting as the data source were not thought of as arbitrary by investigators who were inclined to believe in an elementaristic theoretical model in any case. But when they gave expression to this belief in the limitations they imposed on the data source, they inevitably caused it to produce data that were consonant with their own assumptions.

Data produced under such rules were not permitted to contradict the assumptions on which the rules were based. Instead, it became customary to introduce pseudoexplanations when the social practice of forced numerization produced uncomfortable results. For instance, in investigations of the tactile two-point discrimination threshold so-called paradoxical judgments were common. The subject's skin would be stimulated by either a single point or by two points some distance apart in an attempt to measure tactile sensitivity. Without the help of visual cues, the subject then had to say whether he or she *felt* one or two points. Now, while it was expected that when the two stimulated points were too close together only one point of stimulation would be felt, it was not expected that a single stimulated point would be reported as feeling like two. Yet that is exactly what happened in a significant number of cases.[5] Moreover, the proportion of such "errors" tended to decrease (often quite massively) as the experiments were repeated. This result meant that the numerical value of the sensitivity threshold could vary widely, depending on when in the experimental series it was calculated.

Such results generally led to a search for technical improvements and auxiliary hypotheses rather than to any questioning of the fundamental practices on which these attempts at numerization rested. Reductions in "paradoxical" responses were said to be due to the effects of "practice."

This was hardly an explanation, for it left dark what exactly the word "practice" referred to.

One psychologist who was prepared to probe further, and who achieved an unusual degree of insight into the basis of psychophysical numerizing practices, was Alfred Binet. The first suggestion of what might be going on was provided by his fifteen-year-old daughter, Madeleine, whom he had put through a number of these experiments, only to find the usual "practice" effects. "I knew better this time what the sensations meant," she told him after a later session. "When a sensation was a little 'big,' I thought there must be two points, because it was too thick for one."[6] What Madeleine had picked up was the rather elastic meaning of the category "two" that had to be employed in this situation. Fortunately, her father had no particular stake in the psychophysical methods pioneered by his German rivals and decided to pursue the matter. He extended the investigation to adult subjects, permitting them to report freely on their experiences. Under these conditions Fechner's neat binary scheme of *either one or two* disintegrated. Subjects reported sensations between oneness and twoness – "broad," "thick," and "dumbbell-shaped" sensations that could be categorized as "one" or "two," depending on the subject's interpretation of the schema that the experimenter wished to be applied.[7] Binet concluded that the threshold of two-point discrimination could not be scientifically determined.

But such radical conclusions were the exception rather than the rule. Subsequent investigations did lead to a recognition of the qualitative complexity of the tactile sensations on which the concept of a measurable two-point discrimination threshold had to be imposed. But the constant feature of the various introspective descriptions of the underlying tactile experiences was their *arithmomorphic* form.[8] These experiences always tended to be described in the form of distinct patterns that formed a series between oneness and twoness, such as point, circle, line, dumb-bell, two.[9] Of course, this simply meant that intermediate categories like "point-or-circle" now had to be added, but if one wanted to take the trouble one could go on to determine thresholds between all the adjacent categories in this series. By insisting on distinct categories, and by arranging them in a series, one had already imposed a protoarithmetic form on one's experience, a proceeding that reduced the actual operations of measurement to a purely technical matter. One problem that emerged with the attempt to impose more complex series on the original experience was the lack of agreement among investigators on the number (and quality) of the categories that constituted the series. If one investigator finds five categories between oneness and twoness, and another eight different categories, the arbitrariness of the whole procedure does become rather obvious. The simple two-point threshold did at least have the crucial advantage that a social consensus could be achieved on the basis of the features of the physical instrument of stimulation that was common to all investigators and that indeed had two points.

However, before this physical twoness could be translated into the two-ness of subjective sensation that was the basis of the psychological mea-surement, a little social construction had to be performed. An agreed-upon definition of "two" had to be arrived at by the participants in the inves-tigation. Were tactile sensations that were verbally described as "point-or-circle" or "elongated oval" to be reported as "one" or as "two"? For Madeleine Binet it might simply be a matter of obliging her father, but with adult subjects there was more room for misunderstanding, especially when the usual cultural pre-understanding of the requirements of the sit-uation could not be taken for granted. Thus, when the famous anthro-pological expedition to the Torres Straits added a new dimension to experimental psychological research by attempting to make psychophysical measurements on the local inhabitants, it was found that the average two-point discrimination threshold of the natives was less than half that of Englishmen. The probable reason was guessed by a canny experimenter like Titchener: The English subjects were acting on the basis of their understanding of the norm of accuracy and were refusing to call "two" any sensation that did not indubitably involve two separate points with a distance between them. But the South Sea Islanders, wanting to impress the powerful strangers by their ability to detect the presence of two points, would respond with "two" whenever their actual sensations left any room for doubt.[10]

In the words of E. G. Boring, "the meaning of the judgment *two* is indeterminate unless the criterion has been established."[11] The practice of psychological measurement was clearly grounded in the imposition of agreed-upon definitions on the raw material of individual experience. What was true of the category two in the case of the two-point discrimination threshold was surely true of the other basic judgments, like equal and unequal, on which psychophysical measurement depended.

But perhaps an objectivistic basis for psychological measurement could still be asserted if the measured sensations naturally formed a linear series, so that the socially constructed judgment categories only affected the place in the series where divisions were introduced, but did not affect the serial nature of the experiences themselves. This became the standard defense of orthodox experimentalists, but it hardly convinced their critics. True, psychologists who expected subjective sensations to be measurable were able to offer serially arranged descriptions of the sensations that fell be-tween oneness and twoness, for example. But the skeptics countered with their own introspective examples of the qualitative distinctness of sensa-tions that were supposed to form an incremental series.[12] This became known as the quantity objection to psychophysics, and the discussions that it led to were remarkable for their inconclusiveness.[13]

Both sides to this debate shared a common definition of the problem of psychological measurement. It was seen as *mental* measurement, that is to say, the measurement of entirely private events. If mental measurement in this sense was impossible, then the only conceivable alternative was that what masqueraded under that name was in fact physical measurement. This was a manifestation of the notorious "stimulus error," to which psychologists were said to be prone. But the experimentalists were able to point to their success in establishing relationships between physical and psychological magnitudes, as in Weber's Law. If the sensation side of Fechner's equation was not based on mental but on physical measurement, then Weber's Law would seem to be a law of physics. Yet the basis for the claim that what was being measured were pristine private events was equally dubious. Both sides underestimated the socially constructive, stipulative, aspect of so-called mental measurement. Investigators were indeed able to construct numerizable series of judgments that were different from the operations of physical measurement, although there need be no isomorphic series of private experiences that stood in a one-to-one relationship to the public judgments.[14] Private experience, with its vast potential for interpretation, set few limits to the social constructions that might be imposed on it. The psychological measurements that were the products of this construction certainly had a reality of their own, but it was not a reality that could be neatly classified in terms of the binary categories of the traditional duality of physical body and private mind.

Serializing behavior

Had the meaning of psychological measurement never extended beyond the measurement of private mental events its future would have been extremely limited. Unless one believed, like Fechner, that such measurements could make a contribution to the metaphysical mind–body problem, they were of little theoretical interest. Even the repeated ritual of demonstrating the measurability of mental events must surely have exhausted its appeal after a time. Sensory threshold measurement was of some psychophysiological interest and eventually found practical applications in human engineering. But these successes did not really depend on an answer to the question of whether private mental events were really being measured. What was important in these contexts was the possibility of giving quantitative expression to some *capacity* of the individual organism. Relevant were questions of *performance* against some externally defined criterion, not questions of the quantitative structure of inner sensory experience. Psychological measurement became a generally useful tool only when it was interpreted as constituting a measurement of individual capacity rather than of individual experience. Apart from the jettisoning of philosophical embarrassments, this step enormously expanded the potential scope of psychological measurement, because the number of measurable capacities was limited only by the ingenuity of psychologists in devising tasks with quantified performance criteria.

The problem of psychological measurement therefore shifted from one of imposing a quantitative structure on subjective experience to one of imposing quantitative structure on human action. First of all, if measures of individual action were to function as measures of individual capacity, these actions had to be structured as *performances*. That is to say, they had to result in some product that could be graded against some imposed criterion. Then, in order to give the measurement the appearance of objectivity, an arithmomorphic form had to be imposed on the individual subject's actions. In other words, performances had to be segmented into distinct but equivalent units that could be counted. Under these conditions the numerical value assigned to the performance as a whole seemed to reflect some degree of capacity within the subject rather than the subjective judgment of the investigator.

All the fundamental features of the measurement of psychological capacity were first manifested in Hermann Ebbinghaus's classical work on memory. Using himself as subject, Ebbinghaus endeavored to measure "memory" by constructing lists of nonsense syllables that he presented to himself repeatedly.[15] Whenever he desired to do so, he could attempt to reproduce the list and express the result in terms of the fraction of the original list that he had successfully reproduced. In this way his memorizing performance could be expressed in a number, and he could make quantitative comparisons of his performance under different conditions.

Ebbinghaus was not the only person who was interested in the experimental investigation of human memory around 1880. Wundt, for instance, had some interest in this question and had one of his students, the American K. H. Wolfe, work on the topic. Ebbinghaus and Wundt had one thing in common in that they both reduced the problem of memory to a problem of psychological reproduction, but this common ground makes the difference in their approach all the more striking.[16] In Wundt's laboratory the experimental subject was presented with a standard tone and subsequently with comparison tones at varying intervals of time after the standard tone. In each case the subject had to judge whether the pitch of the standard and the comparison tone was the same or different. The proportion of correct judgments could then be plotted as a function of the time that had elapsed between the presentation of the standard and the comparison tone.[17] What was supposedly being measured here was the ability of the experimental subject to reproduce accurately the sensory memory of the standard tone so that he could compare it with the comparison tone, an ability that seemed to decline with the time that had elapsed since the presentation of the standard tone. For Wundt, psychological reproduction meant the reproduction of a subjective experience,[18] and any attempt at quantification implied a truly mental measurement as it had for Fechner.

For Ebbinghaus, however, psychological reproduction in his famous memory experiments had a rather different meaning. Here it was not a question of whether a particular subjective sensation could be accurately reproduced but of whether a certain objective result could be achieved, irrespective of any private experience. In both cases psychological measurement depended on a matching operation, but in Wundt's case there was an attempt at direct matching of two subjective experiences, whereas for Ebbinghaus it was some public product of the individual's memorizing activity that was matched against the criterion. For Wundt, the objective stimulus was simply the occasion, and the external activity of the subject was simply the expression of the subjective process, which was the real focus of the psychologist's interest. For Ebbinghaus, the matching of the products of the subject's activity with objective requirements was itself the primary focus of experimental interest. The question was how many members of a given series of nonsense syllables could be successfully reproduced or recognized under different conditions. Ebbinghaus's methodology established the category of achieved performance (*Leistung*) as a fundamental organizing principle of experimental psychological research.[19]

This transition entailed certain changes in the nature of the arithmomorphic net that the investigator had to cast over his subject matter in order to come up with a product that could still be called a psychological measurement. It was no longer a question of measuring private experience but of measuring some capacity attributed to an individual, that could be held responsible for a certain level of achieved public performance. That required the juxtaposition of two countable series: the "stimulus" series that defined the task, and a series of individual actions that represented the individual's attempts at measuring up to the task. In Ebbinghaus's case the first of these was represented by his well-known series of nonsense syllables, and the second by a series of repeated attempts at memorizing, or "trials." Let us consider these in turn.

If we merely think of the nonsense syllable as constituting the crucial feature of Ebbinghaus's method, we will have missed its crucial structural feature. He himself was very clear about the fact that his instrument was not the nonsense syllable but the *series* of nonsense syllables. All his experimental questions were addressed to such series, not to individual syllables. (It is worth noting here that the specific experimental tradition to which his work led has been characterized by the serial nature of the stimulus material to which it is bound and not necessarily by the use of the nonsense syllable as such.)[20] All of Ebbinghaus's data were based not on a mere collection of syllables but on a specific arrangement of syllables.

The structure of this arrangement is that of a linear sequence of equivalent units with a determinate length that is countable in terms of the number of units. In fact, he tried using a series of numerical digits for his experimental purposes and only rejected this stimulus material because of some practical problems.[21] At any rate, Ebbinghaus made sure that he had task materials with a quasi-numerical structure, so that their reproduction could lead to a countable number of matches of the constituent elements. With such a structure the counting of matches of individual elements could be considered to give a measure of the adequacy with which memory had reproduced the whole.

But that was only one side of Ebbinghaus's method. The other side involved a structuring of the experimental subject's activity so that the quasi-numerical series, which was characteristic of the stimulus material, recurred on the subjective side of the investigated relationship. The process of the experimental subject's familiarization with the stimulus material was deliberately broken up into a number of equivalent segments, called "trials," which followed one another in a linear chronological sequence. In more traditional investigations experimental subjects might also have to engage the stimulus situation repeatedly, but these repetitions did not constitute a cumulative series. They were not "trials" but repetitions of the experimental observation for the purpose of increasing reliability and decreasing observational error. This aspect also recurs elsewhere in Ebbinghaus's methodology, but of course the point of repeated attempts at memorizing a particular list lies not in mere repetition but in *cumulative* repetition. Ebbinghaus's method is therefore based on the confrontation of two arithmetically structured series, the stimulus series of nonsense syllables and the series of trials that represent the activity of memorizing. This makes it possible to establish functional relationships between what are now two sets of objectively defined variables.

The metaphor of work and the religion of numbers

What induced Ebbinghaus to adopt this approach? He was certainly no behaviorist. It appears that he had certain preconceptions about memory, amounting to a prototheory that suggested to him what sorts of data would be relevant within his conceptual framework. He seems to have started out with a basic conception of memory as a kind of *work*.[22] Instead of thinking of memory as a faculty or being concerned with the reproductive aspect of inner experience, he concentrated on the activity of memorizing, which he saw as an energetic process. The work of memorizing led to an increase in the "strength" of the memory image and made it more easily reproducible. This "strength" of internal representations had nothing to do with their meaning or their quality – it was a purely quantitative, energetic property, and the relation between it and the amount of work invested was conceived by analogy with the corresponding model in the physics of the time.[23] Working within such a framework Ebbinghaus set about constructing a measure of the amount of work invested in the activity of memorizing and a measure of the "strength," or amount of energy accumulated in the memory trace. The former was provided by counting the number of trials, the latter by the proportion of correctly reproduced items in the series of nonsense syllables.

His procedure would have represented no more than a historical curiosity had the metaphor of psychological activity as a kind of work not had such widespread appeal. That appeal was strongest for those psychologists who were interested in making themselves useful in the context of educational institutions. In Germany the tasks facing such psychologists were commonly regarded as calling for an approach to psychological investigation that was fundamentally different from that of the usual laboratory psychology of the time. However, this difference was not seen in terms of a need for statistical information on large numbers of subjects, but in terms of a difference in the focus of investigations with individual subjects. Whereas the laboratory psychology that traced its origins to the laboratories at Leipzig and Göttingen was concerned with the analysis of elementary features of the inner experiences of adults, the new school psychology would have to work with children and would have to be much less restrictive in its focus if it was to produce anything of practical value in the classroom. Yet, school psychology was by no means equivalent to child psychology, which was concerned with general questions of psychological development. The practical tasks of school psychology seemed to require a much more specific focus. That focus was defined by Ernst Meumann, its most prominent German representative, as the child at work.[24] With such an object of study the investigative practices of the new subdiscipline would clearly have to rely very heavily on the pioneering work of Ebbinghaus.

Meumann, who had been Wundt's assistant for six years (1891–1897), was sufficiently steeped in the perspectives of the traditional psychology of consciousness to know that within this framework work could not be regarded as a psychological category. But he thought that when psychological processes were put in the service of "intended or prescribed success" they manifested themselves as psychological work and had to be studied as such. Under these circumstances psychological processes became a means to an end and ought to be studied from the point of view of "psychological economy" – that is, how the desired success might be achieved with the least expenditure of effort.[25] Put in these terms, the investigations of the psychologist were seen to be clearly relevant to the day-to-day interests of educators who wanted to increase the efficiency with which children at school succeeded at their prescribed tasks.[26] More specifically, these investigative interests led to the development of a new field of psychological research concerned with the so-called economy of learning, the

search for "the best means of saving time and energy" in the accomplishment of learning tasks of the type encountered in schools.[27]

The methods most suited to the pursuit of these research goals were
essentially those pioneered by Ebbinghaus. They were based on the use
of two measures, a measure of success at achieving some defined task, and
a measure of the work expended in achieving this success. The former was
most often supplied by the use of serially ordered stimulus material and
the latter invariably involved counting the number of repetitions or "trials"
needed to achieve a certain level of objectively defined success.[28] To use
such measures one had of course to impose the same kind of structure on
the activity of the experimental subjects as had Ebbinghaus. But as this
structure had a great deal in common with the structuring of learning
activities in the schools of the time, no one was likely to regard it as
inappropriate to the limited goals of this type of research.

Limited goals were utterly foreign to Meumann's American counterpart,
Edward Lee Thorndike. Both men led the way in the establishment of a
quantitative psychology of school learning, and on that level the investigative practices they advocated were quite similar. What distinguished them
was the enormous disparity in their influence and the striking divergence
in their knowledge claims.[29] In the history of twentieth-century German
psychology and education, Meumann remains a somewhat marginal figure.
He died relatively young, but much more significant, he lacked both an
institutional powerbase comparable to Thorndike's at Columbia's Teachers
College and a group of socially effective allies like Thorndike's school
superintendents. Meumann's audience consisted mostly of classroom teachers with an interest in rationalizing their work but with little power.[30]
Academic psychologists on the one hand and administrators on the other
saw little or no reason to take much notice of his methods. Thorndike,
however, became the leading representative of an army of educational
efficiency experts[31] and played a central role in the formation of a whole
new branch of psychology that was to supply the rest of the discipline with
many of its basic concepts – the psychology of learning. But for that to
happen his work had to be seen, by himself and by many of his academic
contemporaries and successors, as having a significance far beyond the
walls of the classroom.

For Thorndike, and for the substantial number of American psychologists who shared his outlook, "learning" was not just an activity characteristic of school children but a fundamental biological category applicable
to everything from amoebas to homo sapiens.[32] So the potential scope for
a quantitative investigation of learning was not limited to the special tasks
of a certain kind of educational psychology. The methodology that had
served so well in the narrower context was to be extended to the large
domain of the behavior of organisms in general. This extension was entirely
in line with Thorndike's metaphysics of quantification. "Whatever exists

at all exists in some amount," he proclaimed in his most famous statement, and from investigations that span half a century it is apparent that he equated science with effective measurement of these amounts.[33] Moreover, science as measurement was something worth pursuing with a truly apostolic zeal.[34]

These beliefs were inseparable from the ambitious authority claims of those who represented quantitative science and spoke in its name. These were the experts in social management and control among whom a new breed of psychologist was to occupy a specially important place.[35] Quantitative data by themselves were of course just marks on paper, but they could be transformed into a significant source of social power for those who controlled their production and interpreted their meaning to the non-expert public. Quantitative psychological knowledge was a species of esoteric knowledge that was held to have profound social implications. The keepers of that knowledge were to constitute a new kind of priesthood, which was to replace the traditional philosopher or theologian.[36] As long as the new knowledge was used to buttress the foundations of the prevailing social order – which was invariably the case for Thorndike and those who shared his views – the experts who produced such knowledge were assured of lavish support from the centers of social power.[37] This was the social context in which quantification was gradually transformed from a set of means for the pursuit of specific goals into an end in itself. But on another level quantification did remain a means, a means for the achievement of social status, not only for the discipline as a whole, but for its subareas and for individuals within those subareas. The principles that were successfully invoked for the public legitimation of the discipline always tended to be reflected in the norms that governed the internal status hierarchy within the discipline.

The methodological imperative

Despite some practical difficulties and repeated doubts voiced in various quarters, the commitment of American psychology to quantitative methods was relentless. This was a flood fed from powerful sources. Some of these sources had their origin in the external relations of the discipline, and these will by now be clear enough. Quantification seemed to mark psychology as one of the exact sciences and to distinguish it sharply from such questionable pursuits as philosophy and spiritualism, with which it had been popularly associated. Moreover, as we saw in chapter 7, the first large-scale success of modern psychology as a practically useful discipline had been based on its ability to supply essentially statistical information, and this necessarily placed a premium on quantitative data. The larger legitimation claims of the discipline depended on its ability to make a social contribution that would be regarded as more valuable (i.e., more useful) than that of its potential competitors. Reliance on precise, quantitative data would provide an impressive foundation for such claims.

These powerful forces, generated by the social context within which the discipline had to operate, should not lead us to overlook the importance of the role that quantification came to play in the internal life of the discipline. The most pervasive effects of the social context were to be seen in the elementary fact that psychology changed from a more or less amateurish interest of independent investigators to a collective enterprise organized along disciplinary lines. Thus, it had to develop a set of institutionalized rules that would regulate and coordinate the activity of its individual members so that they contributed, or at least did not visibly harm, the disciplinary project. Rules relating to quantification assumed increasing importance as the discipline grew in size and complexity and in the ambitiousness of its investigative goals.

From its earliest beginnings modern psychology had had to face the paradox that while its *subject matter* was defined as a private object, namely, the individual consciousness, the procedures by which it proposed to study this object presupposed a collective organization, namely, a community of communicating scientists. Their work was to be interdependent and cumulative, as it was in the established natural sciences. Each person's work was to build on that of others, allowing the gradual construction of an edifice of knowledge, brick by brick. That, however, would only be possible if there existed some common recognition of the rules under which this construction was to proceed and some consensus about what constituted a genuine contribution to the edifice of knowledge and what did not. Psychology as a whole had found it peculiarly difficult to achieve agreement on these matters and hence to generate a product that represented *the* truth about its subject matter by common consent. The imposition of a quantitative structure on its knowledge base had seemed to many to offer a resolution of these difficulties. Certainly, quantified items of knowledge seemed to be unambiguous and capable of commanding universal recognition.

Those advantages of quantitative data were however entirely dependent on the possibility of achieving consensus with regard to the rules that governed their production on the one hand, and their interpretation on the other. One might say that the preference for quantitative data merely shifted the problem of achieving consensus onto different, though perhaps more promising, ground. If the ground was indeed more promising, this would presumably be due to the superior prospects, in a numerizing civilization, of gaining adherence to conventions governing the production and interpretation of numerical information.

In the first section of this chapter we explored the fundamental dependence of the process of generating quantitative psychological information on the operation of certain social conventions. Although, as we saw, critical voices were by no means absent when these conventions began to be widely applied, they were quite ineffective in undermining the common faith that what was achieved by adherence to these conventions was more than a merely conventional truth. By the second decade of the twentieth century, the ground rules for the *production* of numerical psychological data were well enough established to be utterly taken for granted. Any debate regarding fundamentals now began to shift to the second major issue, the *interpretation* of quantitative information. It may have been relatively easy, given a particular cultural context, to secure adherence to the conventions that made possible the production of quantitative psychological data, but that still left open the tricky question of what such data meant – in particular, what general inferences one could legitimately draw from such data.

In view of the role that statistical procedures had played in the history of quantitative psychology, it was perhaps inevitable that this question should present itself as a question of *statistical* inference. The Galtonian turn of psychological investigation, discussed in chapters 5 and 7, necessarily multiplied data that were essentially statistical, based on the aggregation of information from many individuals. But even before this development, it had only been possible to detect regularities in the numerized responses of individuals by treating them statistically. This was true of Fechnerian psychophysics as well as of the research tradition begun by Ebbinghaus. Quantitative data in psychology generally meant data that had been statistically arranged. To ask what consensual agreement might be achieved on the basis of such data was therefore easily seen as being equivalent to asking what statistical inferences had a claim to universal assent. This way of putting the question had the advantage of converting a fundamental question about the scientific project of psychology into an essentially technical question that could be solved by the adoption of a few conventions governing the use of certain technical devices. The work of producing quantitative data could then proceed as though there were no unanswered questions about the psychological reality that these data represented.

This pragmatic attitude was clearly the appropriate one in the context of an "applied" psychology that sought answers to practical problems that demanded administrative decisions. Indeed, this is the context in which the attitude can be documented at a very early stage. As outlined in chapter 7, a new era in psychological experimentation had been ushered in by the English school inspector, W. H. Winch, who attempted to improve the efficiency of instruction by comparing the relative improvement of performance in groups of school children subjected to different experimental treatments. There was however one major deficiency in Winch's earliest studies. They were statistically naive and attempted to base general conclusions on a mere comparison of averages. The defect was quickly spotted at University College, London, the fountainhead of the new biometric techniques of statistical analysis. Under the auspices of Charles Spearman, an extensive criticism of Winch's work was soon published in which it was pointed out that averages had to be compared with their "probable errors" before any conclusions could be reached about the "significance" of any difference between them.[38] Statistical techniques for performing this operation were available, so that the question of deciding on the generality of the findings could be reduced to the question of deciding on how many times a difference between means had to exceed the probable error of the difference in order to constitute "true significance." That could be settled by agreeing on a conventional value.

Exemplary pragmatist that he was, Winch immediately saw the point of the criticism and began to adopt the recommended statistical devices in his later studies.[39] Gradually, the estimation of statistical significance became standard practice in the educational psychological literature, replacing the more informal methods of group comparison that had been in vogue previously.[40] Once the practice had become general there was an obvious need to find a shorthand term to refer to the comparison of a statistical difference to an estimate of its variability. That need was filled by the term "critical ratio," which had originally been introduced in the context of comparing the effects of different types of fiscal control of school systems.[41] From the area of educational administration the use of the term quickly spread to the closely related area of educational psychology and thence to general psychology. By the mid–1930s the use of "critical ratios" was an essential feature of much psychological research and merited extended treatment in textbooks of psychological statistics.[42] A search of four prominent journals (*Journal of Experimental Psychology, American Journal of Psychology, Journal of Abnormal and Social Psychology, Journal of Applied Psychology*) indicated that by the period 1934–1936 about one-third of all empirical papers based on group data also reported critical ratios.[43] Most of the rest used correlational techniques.

For practitioners like Winch, such estimates of the significance of differences had made immediate sense because it was believed that they indicated the probability of obtaining a difference as large as the obtained one if the experiment were repeated a large number of times. That was a good basis for making a practical decision about whether it would be worthwhile changing instructional programs on the basis of a limited set of experimental results. However, the early investigators in educational psychology were never content with a simple statement of practical expediency but always claimed that their statistical findings reflected on the nature of fundamental human functions or capacities, like memory or intelligence. Such claims necessarily implied rather more complex issues, for, as the practice of statistical significance testing became widespread in psychological research, the issue that would ultimately have to be faced was that of the relevance of statistical data for psychological theory.

E. G. Boring, the future historian of experimental psychology, appears to have been the first American psychologist to become seriously concerned about this question. It is likely that his concern was part of a more general interest in defending the role of purely scientific psychologists against the spreading influence of applied psychologists within the discipline,[44] but the issues he addressed were real enough. In a series of papers he criticized psychologists' misuse and ignorance of the basis of statistical inference.[45] He wanted to make a distinction between "mathematical" and "scientific" significance, or between statistics and "scientific induction." By the latter he meant the exercise of the investigator's judgment about the psychological significance of a particular statistical finding, a judgment that would have to be made in the light of existing knowledge about the matter that was under investigation. What he deplored was the blind reliance on limited statistical samples, statistical conventions, and statistical assumptions for drawing substantial psychological conclusions: "Statistical ability divorced from a scientific intimacy with the fundamental observations leads nowhere."[46] The danger he sensed was that psychologists were beginning to confuse a constructed statistical reality with the psychological reality they were supposed to be investigating.

His concerns appear to have had little or no discernible influence on the investigative practices of the discipline. He was answered in no uncertain terms by T. L. Kelley, a key figure in the propagation of the idea that psychological research was essentially a statistical enterprise. Boring had summed up his argument as follows: "But, if in psychology we must deal – and it seems we must – with abilities, capacities, dispositions and tendencies, the nature of which we can not accurately define, then it is senseless to seek in the logical process of mathematical elaboration a psychologically significant precision that was not present in the psychological setting of the problem. Just as ignorance will not breed knowledge, so inaccuracy of definition will never yield precision of result."[47] On the contrary, maintained Kelley ("the error of believing that knowledge can be wrought out of ignorance is still strong within me"), that is exactly what a statistical psychology is able to achieve, and he proceeded to demonstrate what he meant by a number of examples taken from the field of mental measurement. His defense of arithmetic units as units of *mental* measurement is interesting: "Without denying the possibility of other workable systems, it seems that, in a civilization such as ours, steeped in the elementary associative and commutative principles of arithmetic and algebra, much is to be gained in simplicity and accuracy of interpretation if the units employed in mental measurement obey these well known laws."[48] What Kelley knew was what a later generation of psychologists was apt to forget –

namely, that the categories of quantitative psychology were intimately dependent on the social practices of a particular culture.

However, for Kelley this dependence was a positive feature, a source of strength. It made it easier to relate the results of psychological research to the practices of social institutions, as when measurement in terms of mental age units could be directly utilized by an age-graded system of elementary education. Beyond this, the adoption of statistical models made it possible to eliminate the element of "intuition" from psychological research and to replace it with strict logical inference. Boring, of course, had argued that this was both a dangerous and an impossible goal, but right from the start it was obvious that he represented a minority opinion. The use of statistical inference spread apace, and as it spread, its use became more rigidly circumscribed by conventions expressed in automatic procedures disseminated through the medium of statistical textbooks.[49]

By the end of the 1930s certain practices in regard to the use of statistics in psychology had become well established. "Statistical significance" had become a widely accepted token of factual status for the products of psychological research. If a particular finding reached a conventionally fixed level of statistical significance, it was automatically received into the corpus of scientific psychological knowledge; if not, it was unlikely to be published, unless accompanied by other, "significant" findings.[50] At the same time, there was virtually no appreciation of the fact that commonly used statistical methods implied specific models of reality when applied to numerical data with psychological reference. These methods were looked upon as theory-free or, at a minimum, theory-neutral techniques whose employment was unproblematical, once a few conventions governing their use had been fixed. The commonly used textbooks of psychological statistics encouraged these attitudes by a cookbook approach to the subject, which avoided discussion of fundamental and controversial issues. In most cases the methods were presented anonymously, without human attribution, thus imparting to them a distinct aura of infallibility.[51]

These developments had certain undeniable advantages for the internal life of the discipline. In the first half-century of its existence, modern psychology had failed to generate any body of theory that was simultaneously precise, powerful (in the sense of wide applicability), and generally acceptable to most workers in the field. Yet, investigators continued to produce ambiguous data that were claimed to be scientific. That claim was difficult to justify on rational, theoretical grounds, given the lack of consensus on that level. The only alternative basis for that modicum of agreement that was necessary for saving the scientific status of the field was to be found on the level of methodology.[52] Purely technical criteria could be used as a substitute for theoretical or substantive criteria in judging the value, and even the acceptability, of the products of psychological investigation. The ideal that began to emerge toward the end of our period was that of the automatization of the process of scientific inference. Although it might not be held together by substantial theoretical commitments a diverse group of investigators might still be able to function as a community of sorts if it adopted a specific set of unexamined methodological conventions.

One such convention that deserves special mention relates to the use of average data. As we saw in chapter 5, the averaging of quantitative information from many subjects was not part of the classical methodology of experimental psychology. However, with the demonstration of the usefulness of statistical data for many of the practical contexts in which psychology was beginning to be applied, such data gained a certain respect. When experimental methods were extended to the investigation of complex, molar behavior patterns – for which they had not originally been considered suitable – the use of average data seemed to provide an acceptable way of coping with the utter lack of consistency that was characteristic of complex individual behavior forced into the quantitative mold. Individual behavior might show little or no consistency from one occasion to another or from one experimental situation to another, but if data from many individuals were pooled, certain statistical regularities sometimes emerged. These regularities were to form the basis of psychological generalizations, even though the average pattern might not correspond to the actual behavior of a single individual member of the statistical group.[53]

This trend was particularly noticeable in the new field of the psychology of "learning," understood now in the very broad sense that men like Thorndike had given to the term and that had been adopted by functionalists like Carr as well as by behaviorists. The abstract psychological category of "learning" was intended to lead to the formulation of general laws of adaptive behavior that held for all animal organisms under all natural and artificial conditions. This was the field that, it was hoped, would provide a psychology by now strongly tied to the goal of behavior control, with the fundamental science on which its practical prescriptions would be based. A trickle of empirical studies of human and animal learning soon became a flood, the maze quickly establishing itself as the preferred physical setting for these investigations. However, it was not long before the gross inconsistency of individual performances in these experimental settings began to raise some doubts. A protracted discussion in the pages of the *Journal of Comparative Psychology* seems to have settled the issue for most of those working in this field.[54] Henceforth, the use of average or group data was to become the norm for this prestigious area of basic research. By 1935 an editorial statement in the same journal had to enter a gentle caveat against the pervasive use of "the average animal."[55] Nevertheless, reliance on this fictional creature became even more pronounced. Finally, it achieved the status of a methodological fetish, and the minority who refused to worship at its feet were forced to establish their own publication outlet, well out of the mainstream.[56]

Two things are striking about this development. The first concerns the confusion between statistical issues and substantive psychological issues, which was present from the first, and the second concerns the ease with which statistical significance testing became accepted as an indispensible tool for the corroboration of psychological theories. These two features are not unrelated. What is noteworthy about the original confrontation of the issue of group data in the *Journal of Comparative Psychology* is that the issue is already defined as a statistical problem and only as a statistical problem. It is seen as a matter of handling intraindividual and interindividual *variability* on predetermined parameters. The question of the psychological appropriateness of these parameters and of the *psychological* significance of the measured variability never comes up. In particular, the requirement of the statistical model – that one deal with *continuous* magnitudes – was blindly accepted as binding on the underlying psychological reality. Thus, the possibility that the observed inconsistencies might be at least partly the result of underlying psychological discontinuities was not an alternative hypothesis that was available to these investigators.[57] For them there was no disjunction between the properties of the statistical aggregates they had constructed and the distribution of those psychological characteristics that had been drawn on to produce these artifacts. The problem of variability could therefore be reduced to one of how to test for *statistical* significance, conveniently ignoring the crucial question of what restrictions the statistical model implied by this practice imposed on the concepts of psychology.[58]

Establishing the statistical significance of differences among aggregates was a practice of proven value in practical contexts where the overall effects of administrative decisions needed to be assessed. Given the prevailing tendency to reify statistical artifacts, and therefore to confuse statistical with psychological reality, it was quite natural for statistical significance testing to be employed as a basis for decisions about the validity of *psychological* hypotheses. Among the factors that helped to spread this practice, the automatization of scientific decision making, which it encouraged, has already been mentioned. More generally, it was a practice that reduced the demands made on psychological theorizing – no trivial achievement for a discipline that had never been able to get its theoretical house in order. Using statistical significance tests as the standard technique for the corroboration of psychological hypotheses meant that theories were generally regarded as "confirmed" if a very weak logical complement, "chance," or "the null hypothesis" could be disconfirmed.[59] This travesty of scientific method certainly allowed the growth of a major research industry that offered employment to many, even though its products can now be seen as having contributed nothing of either practical or theoretical value.[60]

Not only did these methodological aberrations legitimate a large-scale waste of time, effort, and resources, they also confined psychological theorizing in an increasingly narrow mold, thus closing the door on alternative conceptualizations and practices that might have reversed the process of intellectual decline that the discipline was now beginning to suffer. Because methods were almost universally regarded as theory-neutral instruments, the assumptions on which they were based were able to function as unrecognized theoretical models for predefining basic features of the reality under investigation.[61] The more rigidly the demands of a particular statistical methodology were enforced, the more effectively were ideas that did not fit the underlying model removed from serious consideration. Such ideas had first to be translated into a theoretical language that conformed to the reigning model before they could be seriously considered. In other words, they had to be eviscerated to the point where they no longer constituted a threat to the dominant system of preconceptions guiding investigative practices. The final stage of this process was reached when the statistical models on which psychologists had based their own practice were duplicated in their theories about human cognition in general.[62]

One of the fundamental sociopsychological principles described by Wilhelm Wundt was based on what he called the "heterogony of ends."[63] By this he meant the tendency for human goals to have unintended consequences, which then modify the goals themselves. Such developments he tended to see as progress. The history of the discipline for whose paternity he was often held responsible certainly provided some striking illustrations of this principle. Fortunately for him he was spared the difficult task of reconciling the resulting developments with his rather touching nineteenth-century belief in the inevitability of progress.

10

Investigating persons

Extension of the mental test model

American psychologists, as we noted in chapter 7, had set about redefining the psychological problem of individual differences in terms of a comparison of performances against a competitively defined standard. This redefinition appeared to be perfectly adapted to the administrative requirements of social institutions whose efficient functioning depended upon some kind of selection process that was rationally defensible. The kinds of tasks that best lent themselves to the measurement of individual differences in this sense were tasks with answers that were unambiguously right or wrong. Performance on such tasks could be readily arithmetized by counting right and wrong answers, a procedure that had long been conventionalized in the institution of school examinations. Improvements in the precision of such procedures, made possible by Galton's fundamental contributions, eventually led to the construction of normed intelligence tests.

It is hardly a cause for surprise that educational institutions provided the setting in which intelligence tests proved themselves best adapted to administrative requirements. Transferred outside their home ground, their ability to predict future performance was much diminished, often to the point where their practical usefulness vanished altogether. This problem already emerged in the first major attempt at employing intelligence tests in the service of a noneducational institution, the U.S. Army in World War I. Most of the psychologists involved appear to have been convinced of the potential applicability of intelligence tests to military selection problems. However, although they were given considerable leeway and succeeded in testing well over a million men, the army does not seem to have found the results of this exercise particularly useful.[1] The military administrators were apparently more impressed by the contributions of a second group of psychologists who were prepared to tackle selection problems by means of an evaluation of nonintellectual qualities like loyalty, stability, vigor, and regard for authority. This second group worked under the direction of Walter Dill Scott, the only psychologist to be awarded a military decoration for his professional services to the army.[2]

Scott's academic base was in a business school where he had been addressing himself for some time to the selection problems faced by private employers of white collar workers. In particular, he had developed techniques for the selection of effective salesmen, and he recognized the potential applicability of these techniques in other institutional contexts – for instance, when looking for leadership qualities in the selection of officer candidates. In such practical contexts the ability to solve quasi-scholastic problems with unambiguously right or wrong answers obviously counted for little. What did count were personal qualities of a generally nonintellectual kind. The problem was to find a way of attaching numerical values to such qualities, so that individuals could be quantitatively compared with one another for the purpose of making the appropriate selection decisions. Scott's solution was to analyze the task requirements into a number of components and then to relate these components to designated personal qualities. Thus, a salesman's effectiveness was thought to depend on the components of "appearance and manner" and of "conversational effectiveness," while his value to the firm depended on "character," among other components. Any candidate could then be given a numerical rating on each of these components by comparing him to a standard group known to the interviewer. In making such assessments the rater was explicitly asked to take into account specific personal qualities such as cheerfulness and self-confidence in rating "appearance and manner," and honesty and ambition in rating "character." Much the same approach was used in officer selection.

Here then was a first approach to the quantification of personality. The possibility of quantification arose out of the fact that certain socially defined tasks could be carried out with different degrees of success by different individuals. As it was believed that these differences in task accomplishment depended on specific personal qualities, the quantitative prediction of degree of success in the task was automatically linked to an assessment of these personal qualities. At first this assessment remained embedded in a specific practical context. But the implicit model of human personality with which practical psychologists like Scott operated had obvious attractions for psychologists who were interested in developing a natural science of the universal characteristics of human individuals. In the implicit model, which underlay the rating practices of personnel selection, the individual was regarded as a collection of discrete, stable, and general qualities. These qualities were thought of as being identical in kind from one individual to another, varying only in degree. Moreover, they were qualities that were manifest in assessment situations as well as in criterion situations for which prediction was desired. It was not long before the potentialities of this model were recognized and exploited.

During the 1920s it became more and more apparent that the early hopes for the practical efficacy of intelligence tests would not be fulfilled. Within the school system, their major area of application, these tests had proved capable of meeting administrative needs only up to a point. They did show some correlation with academic performance but not impressively so. The prediction of academic performance from intelligence-test results remained highly approximate, and this gave rise to a search for ways of closing the gap between predictive ambitions and the reality. A consensus quickly developed that intelligence was only one of the determinants of real-life performance, and that the other determinants would have to be looked for under the rubric of "character" or "personality." As a writer in the journal *Educational Administration and Supervision* put it: "It is quite probable that we have gone as far as we can go in psychology in predicting future success on the basis of things intellectual. The next development in this field must come through experimental study of character and personality traits. That character, personality, attitude, or whatever we may choose to call it, plays an important part in success cannot be doubted."[3]

This was not an isolated opinion. The literature of the period is full of similar suggestions, and in a number of cases these suggestions are immediately translated into practice.[4] What is noteworthy about the expert opinion of the time is that the limited success of mental testing leads neither to a questioning of the administrative goals of testing nor to a questioning of its fundamental methodology. It is taken for granted that the goal of mental testing is the prediction of individual success in terms of administratively defined criteria. Intelligence tests had brought the achievement of this goal somewhat nearer. If they fell short of complete success, the fault was not likely to lie in any fundamental inappropriateness of the method; it was suggested, rather, that the domain of application of the method had not been sufficiently extensive. What was measured by intelligence tests, it now seemed, was not really a global mental power, or overall adaptive prowess, as the eugenicists among the testers had wanted to believe, but a more circumscribed kind of potential of a more specifically intellectual character.[5] For more effective practical prediction the measurement of intelligence had to be complemented by the measurement of nonintellectual characteristics, which intelligence tests did not tap.

During this period the technology of intelligence testing was a source of pride to many American psychologists. It had been crucial in transforming the discipline from an ivory-tower pursuit to a significant form of social practice. Small wonder that it served as a model of technical sophistication when the problem of measuring nonintellectual features began to be faced on a large scale. There had been earlier attempts at developing tests in this area, the association test and Downey's will–temperament test in particular. But these were now eclipsed by other instruments better adapted to the requirements of what was beginning to be called "personality measurement."[6] The new instruments copied the features that had made group intelligence testing such an economical proposition. They were paper-and-pencil tests of simple format that could be readily administered to large groups of subjects and scored quickly and without any need for thought. Personality rating scales and inventories began to proliferate.

Constructing "personality" as an object of research

Two features of these instruments stand out: their reliance on an additive model of the human person and their identification of conventional verbal description with psychological reality. Such instruments consisted of a verbal label or title, like "ascendance" or "introversion," and a numerical structure implying that the quality referred to by the label existed in some measurable amount. The units for the operation of measurement were supplied either by the arbitrary subdivisions on a rating scale, or by the verbal items constituting a personality inventory. In either case, the psychological reality that was being constructed here involved the positing of features that were independently definable and isolable and that retained their identity across situations and across persons. Thus, the "introversion" in terms of which the behavior of individual A in situation X was described was assumed to be entirely comparable, not only with the "introversion" ascribed to the behavior of individual A in situation Y, but also with the "introversion" attributed to individual B in situation Y. Any difference in B's "introversion" in situation Y and A's "introversion" in situation X was purely quantitative. Qualitative variation of features had been extracted both from the description of individuals and from interindividual comparison and had been relegated to the abstract level of trait descriptions.[7] One abstract trait differed qualitatively from other abstract traits, but individuals were limited to possessing larger or smaller quantities of identical features. These quantities were arrived at empirically by additive composition of items and comparison of the individual "raw scores" so generated.

It is instructive to compare this procedure with that which had been so successfully employed in the construction of measures of intellectual capacity. In that case the effectiveness of individuals had been gauged in task situations to which an unambiguous performance criterion could be applied. An answer was either right or wrong, and a larger number of right answers could be held to reflect a greater quantity of some ability. The use of task performance for the construction of scales of ability depended on the elimination of ambiguity. That is why only the final answer to a question, or "intellectual yield," could be considered, and differences in the way that answer was arrived at had to be ignored.

In the construction of instruments for the measurement of ability, the elimination of ambiguity was thus achieved by restricting permissible tasks to those with unambiguously right or wrong answers and by ignoring the actual intellectual processes that produced a particular result. In constructing instruments for personality measurement, an analogous elimination of ambiguity had to be performed. This depended on two crucial operations, which bore some resemblance to those that had been employed in the construction of devices for the measurement of ability. The place of tasks with unambiguously right or wrong answers was taken by verbally labeled categories that retained an absolutely identical meaning irrespective of personal or situational context. This made it possible to carry out unambiguous attributions of quantities of the same characteristic across individuals and across situations in the manner already discussed.

The second potential source of ambiguity in personality assessment had its origin in the subjective processes that led up to the final rating. The meaning of a particular rating will depend on the connotations that the verbal trait label (or the key words in a questionnaire item) have acquired for the rater. It will also depend on the rater's sensitivity to the nuances of other people's conduct, on the rater's implicit psychological theories, on his or her preference for certain kinds of judgment, and on a host of other subjective factors. The new technique of personality assessment had a simple way around these sources of ambiguity by stipulating at the outset that the results of personality ratings were to be interpreted as reflecting attributes of the target person rather than attributes of the rater.[8] Even in self-ratings, when target and rater were the same person, the results of the exercise were held to reflect on that person as the object and not as the subject of this self-reflective action.

The arbitrary element in this construal becomes quite evident when we compare the meaning attributed to the results of intelligence tests with that attributed to the results of personality ratings. In both cases we have the results of a task performance that are a joint function of the task and of the person engaged in it. In the case of personality ratings the task is to reduce the complex input received from the target person to a series of marks on simple scales. In principle, the outcome of such a task situation could be regarded as a reflection of some attribute of the rater or some attribute of the task. For instance, one could interpret individual differences in rating the same target(s) as a measure of the raters' sensitivity in applying particular judgment categories. But personality-rating tasks were typically not used to make inferences about the psychology of judgment. During the early period of personality psychology, and to a considerable extent thereafter, it was simply assumed that personality ratings were an unproblematic product of attributes of the task, not attributes of the rater.

In particular, such ratings were held to be a direct reflection of that part of the task that was constituted by the input received from the target person.

In the case of ability tests, any particular outcome can also be regarded as depending on both the person tested and the tasks. Just as persons can be ranked on the attribute of ability, tasks can be ranked on the attribute of difficulty. Does an individual fail at a particular task because the task is too difficult or because the individual lacks ability? The preferred explanation depended on prior theoretical commitments, like the belief in a fixed and inherited "general ability," and on the social interests behind those commitments.

Both intelligence testing and personality rating relied on problem-solving situations for their basic data. But by an act of constructive preselection the object of investigation was determined as being entirely different in the two cases. In the one case, performances in the problem-solving situation were held to reflect some inherent property of the problem solver (i.e., intelligence); in the other case, performances were taken to reflect inherent properties of the task (i.e., personality characteristics of the target person). The crucial point here is that the fundamental psychological meaning and reference of the empirical data were constituted by an interpretive construction that was not derived from those data but preceeded their collection.

With a Galtonian paradigm of research practice, the arbitrary nature of the interpretative framework was typically disguised by operations of selection that limited the available empirical material to data appropriate to the preselected interpretive pole. Such data consisted of variances. If one wanted an empirical construction that lent itself to an interpretation in terms of differences in individual ability, one sampled across individual problem solvers and worked with interindividual variances on the same set of tasks. But if the task was one of personality rating, one produced sets of variances between the rated individuals rather than the raters. In either case the unwanted interpretation would become empirically invisible because there would be no corresponding variance to be interpreted. Of course, investigators were aware that in any particular case the outcome of the performance depended both on the performer and on the task object. Their split investigative practice made sense as long as one quietly assumed an additive relationship between these aspects, so that the one could be studied independently of the other. For that assumption, however, there was no warrant other than the investigative practices themselves.

In the course of time differences among raters were given some recognition in the form of interrater reliabilities. Once again, substantive problems were sidestepped by replacing them with a set of interindividual variances. Had these substantive problems been pursued, one crucial reason for differences among raters would have been traced to the fundamental dependence of the method on ambiguous natural language terms. This was the case, both in the use of global trait categories, like "ascendance–submission," and in the interpretation of specific items in an inventory, all of which relied on the raters' natural language competence. Personality ratings could very easily be recast as tests of the raters' sensitivity and competence in applying certain categories of their language. But of course, the ratings were supposed to be a direct reflection of the target personality, not a test of the raters' competence as members of a particular language community. It was as if intelligence-test results had been held to provide only a measure of some objective property of the tasks and had never been related to the ability of those being tested.

Although personality ratings and inventories were essentially an exercise in the application of certain verbal categories, they were presented as somehow analogous to natural scientific measurement. A particular set of natural language terms could therefore be made to take on the guise of categories of nature. The question of how terms like "ascendance" or "dependence" functioned in the language games characterizing certain social relationships was not the kind of question that motivated these investigative practices. Instead, "ascendance," "dependence," and the rest, were treated as unambiguous properties of the natural world that were to be investigated much as a nineteenth-century physicist might have investigated electrical resistance. What this amounted to was a masquerade in which categories generated by a very specific social order were held to represent an ahistorical natural order.

The categories with which the new psychology of personality operated were always highly dependent on a particular social context. This was quite evident when its characteristic investigative procedures first appeared in the form of personal ratings in the practice of personnel selection. Obviously, the categories on which prospective salesmen or prospective officers were rated depended on what was considered to be relevant to the efficient performance of their jobs. The social context in which the rating procedure was applied automatically provided criteria for the selection of a few personal qualities out of the potentially unmanageable multiplicity of such qualities. However, when personality psychology began to be pursued independently of such specific practical settings, it did not mean that it now selected its categories independently of any social context. No psychological theory functioned as a source of criteria for deciding which, out of a virtually unlimited variety of possible trait concepts, were to be regarded as particularly important. In the absence of any theoretical model that might distinguish among traits in terms of their *psychological* significance, the choice of the personality categories that formed the object of investigative practice was ultimately determined by their apparent practical usefulness or by their "intuitive" interest, the latter entirely governed by cultural factors that were subject to the usual limitations of time and place.[9]

One consequence of this was an extraordinary fluctuation in the personality categories considered worthy of psychological investigation at different times. Thus, if one examines all the articles published in a key journal, like the *Journal of Abnormal and Social Psychology,* over a period of several decades, one finds considerable variation in the relative frequency with which various terms referring to personality traits are used.[10] Comparing the period 1920–1935 with the period 1940–1955, there is a sharp decline in the popularity of the early favorites: The personality category of "extraversion," for example, drops from being used in 39 percent of all articles employing trait terms to only 6 percent in the later period, and "ascendance" drops from 21 to 3 percent. On the other hand, the category of "emotionality" rises from 11 to 22 percent. Some categories disappear in the course of time, others appear and flourish for a while, only to be eclipsed in their turn. There appear to be no grounds intrinsic to the subject matter for this constantly shifting empirical basis of trait psychology. It is much more likely that these changes represent the coming and going of research fads, but in this case the fads are directly related to events in the social environment of the discipline.

Thus, the strong attention paid to the polar categories of "ascendance" and "submission" during the early period of American personality research appears to reflect its practical roots in personnel and officer selection. In both those contexts the goal was to pick individuals who could be relied upon by their superiors to take charge of situations and impose their intentions on others. This was conceptualized in terms of stable intraindividual qualities that varied in amount from one person to another. Psychologists not only contributed a technique that could be claimed to measure this amount but also supplied a reified category that translated the general social requirement into psychological terms. Numerous instances of the socially desired outcome were aggregated to contribute to an abstract quality of "ascendance," or its opposite, that was now thought to reside not in interpersonal situations but within strictly autonomous individuals. This made it possible to conceive of a new field of research that would take such general intraindividual qualities as its object and that would not be directly tied to specific practical requirements.

The construction of personality traits as objects of investigation led to the emergence of a brand new area of psychological research. In the American Psychological Association Yearbook for 1918, not a single psychologist listed "personality" as a research interest, but by 1937, 7 percent did so.[11] The earliest attempts to review the new research area already show that it was dominated by the trait concept.[12] It was the investigative practice of trait measurement that was to translate certain culturally embedded concerns into something that could sail under the flag of science.

The uses of personality research

Personality research would hardly have achieved the kind of position it did without an ideological component that gave a more general significance to its rather modest technical roots in personnel selection. During the founding period of American personality research, this component was supplied by a powerful and well-funded social movement, the mental hygiene movement. Growing out of seeds sown in the period before World War I, this movement achieved a pervasive influence during the 1920s and 1930s.[13] At first confined to more specifically psychiatric concerns, it quickly spread to the educational system and to the area of child care in general. It affected many of the institutions with which psychologists had practical links and created new ones that psychologists were often able to exploit for the advancement of their own professional project. Favored by generous grants from private foundations,[14] the mental hygiene movement was based on the twin beliefs that the causes of social and interpersonal problems were to be located in the maladjustment of individuals and that the origins of such maladjustment lay in the histories of individuals – that is, in childhood. Interpreting social life in terms of metaphors of health and illness, the mental hygiene movement projected hopes of a better future that was to emerge, not through the conflict of collective social interests, but through the "treatment" of individual maladjustment by the appropriate agencies of social control.[15] If such treatment was usually ineffectual, this could only be because it lacked a scientific basis, because not enough was known of the causes of individual maladjustment. A better knowledge base was needed, and psychologists were just the people to produce it.

Two areas of psychological research were primarily affected by the mental hygiene movement, child development and personality research. The effect on the former was quite direct, taking the form of massive grants of money.[16] In the case of the very new area of personality research the effect was more subtle but nevertheless beneficial in the long run. The concept of "personality" played a central role in the ideology of mental hygiene. "Personality" was posited as the object, which was to be the focus of a variety of energetically pursued programs of intervention in the lives of individuals.[17] In practice, these programs were quite diverse but they were thought of as converging on a common object, namely "personality."[18] Psychiatrists, social workers, and educationists occupied the key positions in the mental hygiene movement because of their activist involvement in practical programs, but psychologists found their own niche by functioning as a source of expert knowledge about the common object of these programs. Their claims were based on their supposed ability to marshall the techniques of science in the service of the common goal. In practice this generally came down to the construction and application of scales that would subject "personality" to the rigors of measurement and so convert it from merely an object of social intervention into an object of science. Their transformation of a presupposition for common practical action into a reified natural object was no mean service.

For the discipline of psychology the new field provided an opportunity for achieving another felicitous synthesis of the demands of practical relevance and scientific appearances. But the costs of this achievement were considerable. What was given up was an earlier tradition of detailed study of individuals and of specific life events. When the *Journal of Abnormal and Social Psychology* was taken over by the American Psychological Association in 1925, there was an immediate and dramatic change in the investigative practices on which published articles were based. Whereas in 1924 80 percent of the empirical papers published were based on the study of individual cases, that proportion dropped to 25 percent in the following year. Their place was taken by statistical studies based on group data, as was required by the prevailing Galtonian paradigm for psychological measurement.[19] Individuals were to be known only through their standing in a group. The display of numbers was able to pass for science, and the information produced by these means had its practical uses in the context of mass programs of therapeutic intervention. But the kind of knowledge of individuals that was generated in this way always remained a knowledge of strangers.[20] "Personality" as a constructed object of these investigations never had anything in common with traditional concepts of the human person as a social agent.

Character as an investigative object in Germany

In 1933 the British psychologist, P. E. Vernon, published an even-handed critical comment on the extreme contrast between American and German approaches to the study of personality.[21] American investigators, in their almost exclusive reliance on tests, seemed to have substituted technology for science. Conventional European opinion, Vernon thought, would probably dismiss the testing approach as arising from the desire to exploit and improve people, rather than to understand people. In Europe, the study of personality was particularly well developed in Germany, and there the qualitative approach ruled supreme. From this point of view it made no sense to aggregate trait values from different individuals because the quality of a trait was part of the whole personality and therefore would vary from one person to another. However, when it came to validating their personality descriptions, the methods of the German investigators were "unashamedly subjective."

These methods, which Vernon dismissed in two words, actually represent a rather interesting chapter in the history of psychology's investigative practices precisely because the contrast between them and the dominant practices is so extreme. Their "subjectivity" was anything but naive, and they were taken very seriously by many European psychologists at the time. How was such an extreme divergence possible within the same discipline? To answer this question, we will have to begin by describing the basis of these practices and then take up the question of the social conditions under which they flourished.

Despite their diversity, all the practices we have considered so far have in common a certain kind of relationship between the investigator and the object of his or her investigation. This relationship could be described as distanced and as intellectual. A major point of studying people in special situations like experiments or tests was to preclude the development of a normal human interaction among subjects and investigators. The latter, in particular, were to treat their human reactions to their experimental subjects as scientifically irrelevant and were to regard them simply as objects of intellectual interest. Any reaction that an investigator had to the subject as a human person was definitely not part of the official definition of an investigative situation. The investigator figured in this definition purely as a machine for analyzing data and drawing logical conclusions.

Even before the end of the nineteenth century there were those who felt that this feature of the new psychology precluded its application to the study of personality. By far the most influential voice was that of Wilhelm Dilthey, who sharply expressed the general distaste that practitioners of the *Geisteswissenschaften* felt toward an experimental psychology. Such a psychology, according to Dilthey,[22] would be of no use to the historian, for example, who had to make sense of the actions of individual personalities under specific historical circumstances. The results of experimental psychology would be irrelevant in this context because, at best, they might lead to an "explanation" of human behavior that was analogous to a physicist's explanation for the behavior of an inanimate object. Experimental, or "explanatory," psychology was only able to achieve its results by treating human individuals as objects – that is, by pretending that our way of knowing other people was not fundamentally different from our way of knowing planets or plants. But everybody knows that that is not true in everyday life. We accept other people as having experiences, feelings, and desires, just like ourselves, and this means that their actions are directly meaningful to us. Dilthey's proposal was that this everyday experience be accepted as the basis for a different kind of psychology. He called it *Verstehend* (understanding) because its aim was not the production of covering law-type explanations but the elucidation of the psychological meaning that could be directly sensed in a person's actions.

Dilthey's proposal went rather deep. It implied a departure from the entire Enlightenment tradition of knowledge production, according to which only rationally processed sensory knowledge could provide the basis for a science. For Dilthey, however, the basis of all knowledge was not any kind of neutral, gutless, elementary sense experience, but something for which he and his many followers used the term *Erlebnis*. The closest English equivalent is "experience," but more in the sense of "living through" or "witnessing" than in the sense of dispassionate observation.

In an interpersonal context this manifested itself as a direct intuiting of the pattern of meaning running through another individual's actions or the products of those actions.

Other German writers developed these ideas in a direction that led to the emergence of a new kind of psychological practice. Theodor Lipps, who was much more involved in psychology than Dilthey, gave a more concrete turn to the latter's rather general suggestions. Lipps was of the opinion that we have a general tendency to reproduce, at least in imagination, the affectively toned actions and gestures of others.[23] Because the other person's meaning is expressed in these gestures, we are able to experience *(erleben)* that meaning ourselves by reproducing the gestures.[24] This is the basis for a direct psychological understanding of other people.[25]

Ludwig Klages gave a similar, though less specific, account of the way in which we understand other people. In his case, however, this general orientation is very much linked to a specific, practical methodology, namely, the interpretation of handwriting.[26] Klages's approach to graphology is definitely to be referred to as an *interpretation* rather than an *analysis* of handwriting. Unlike his predecessors in this area,[27] he did not attempt to correlate specific features of handwriting with specific personality traits. Such a procedure would have introduced nothing new into the investigative practices of psychology. Klages, however, advocated a procedure whereby the psychological significance of each handwriting feature would have to be determined in the light of the overall impression that the handwriting made on the graphologist. The latter was supposed to be able to reproduce in herself, and thus to know directly, the overall quality of experience that the handwriting expressed, whether it was rich, deep, and warm or restricted, flat, and cold.

It is obvious that a different kind of psychological practice has been introduced here. The separation between the investigator as subject and the external object of investigation has become fuzzy. The investigator's affective reactions have become an important component of investigative practice, and the personal characteristics of the investigator have become crucial. Other methods were based on objectively specifiable procedures that could be more or less automatized, but in the new graphology the intuitive talent of the individual graphologist constituted a basic and irreplaceable factor.

Although graphology was the area in which the new approach to psychological research was worked out most systematically, it was not the only area that provided a stage for a practice based on the principle of *erleben*. Other techniques, which had for the most part had a long and decidedly checkered history,[28] were recruited in the service of the new approach. These techniques were grouped together under the heading of "the psychology of expression," and they included the interpretation of facial expression, gesture, posture and movement, voice quality, and even body build and physiognomic appearance. All these were features that were open to psychological interpretation if one thought of them as directly expressing the quality of the individual's affective life and personality. These direct expressions could then be picked up and characterized by talented and experienced interpreters.

The skill of the individual interpreter might be considered crucial, but this did not mean that he or she was given a free hand with regard to the terms in which the interpretation was to be expressed. In principle, the interpretation might have been framed in terms of episodic psychological events or situational factors. For instance, facial expression might have been of interest because of its relation to fluctuating affective experiences, or posture might have been studied in relation to the social psychological situation.[29] But these were not the knowledge objects that these German students of personality favored. The object that they sought to construct by means of their expressive methods was categorized as "character,"[30] a concept that represented an objectification and a synthesis of a number of rules for the explanation of human conduct.

Among the more important of these rules were the following: Human conduct was not to be explained as situationally determined but as issuing from sources entirely within the individual; these inner sources were not to be regarded as changing and developing but were to be thought of as static, permanent and unchangeable structures; the inner sources of human conduct were not to be described in a neutral, quasi-naturalistic language because they had an inherent value, positive or negative, and were therefore to be referred to in ways that expressed this value; the inner sources of human conduct were always to be thought of as constituting a unity, not as an aggregate of components.

The very term "psychology of expression" or "expressive methods" already presupposed a particular model of psychological explanation, incorporating at least some of the above rules. By defining certain features of human behavior as "expressive," these psychologists could consider themselves to be studying not the behavior as such but the presupposed underlying characterological reality. Moreover, the philosophy of *erleben* gave them reason to believe that they were able to have direct access to this reality. The norm of separating factual statements from theoretical statements, characteristic of natural science, would not apply; instead, empirical and theoretical statements were fused in an analogical synthesis. A specimen of handwriting, for instance, might be described as cramped or expansive, but this would simultaneously function as the description of an underlying personal reality. There could be no question, with such a procedure, of "validating" one's theoretical claims – they were indissolubly fused with relatively empirical statements.

One might well be inclined to wonder what kind of knowledge interest such a form of psychological practice could conceivably serve. For some of the early practitioners the explanation seems to lie in the persisting dependence of German psychology on German philosophy, already referred to in chapter 8. German psychologists were not only under pressure to legitimize their contribution in the eyes of philosophers in institutionalized positions of authority; they often owed their academic appointments to their judged competence in the field of philosophy.[31] When the irrationalist wing within German philosophy gained in influence, as it did in the early part of the twentieth century,[32] this could hardly fail to have its effects on at least some German psychologists.

But such factors would hardly have sufficed to transform a romantic fantasy into a historically significant form of psychological practice. The crucial impetus for this development came from outside the academic world. After World War I there was an, albeit limited, market in German industry for procedures that claimed to assess not only individuals' technical skills but also their personal qualities.[33] Graphology, in particular, was able to meet the requirements of this market. But this hardly accounts for the general efflorescence of expressive methods, especially in the 1930s. To discover the crucial factor in this development, we must look in another direction. It is a direction from which American psychology had previously received a crucial boost, namely the military.

Whereas American psychologists were able to turn their army work during World War I to great advantage for the prestige of their discipline, this had not been the case in Germany. There, the psychological contribution during World War I had been limited to the selection of a few specialists like radio telegraphists and artillery spotters. The big chance for the collaboration of German psychology with the German military began a decade later, during the Weimar Republic. Under the Treaty of Versailles, the German officer corps had been greatly reduced in size, which had the effect of making the military authorities determined to make up in quality at least part of what they lacked in quantity. In the immediate postwar period the persistence of the old military caste system and people's recent memories of the horrors of war prevented any innovations. But a decade later a military career was once again attractive to many. There was now a large pool of applicants for a very small number of officer vacancies.[34] At the same time, political criticism of the traditional recruitment of officers from a privileged social stratum began to become quite vocal. The introduction of personality tests for all officer candidates offered a solution to both problems. It established a basis for officer selection that could replace family status, and it provided an apparently nonarbitrary procedure for selecting the most promising candidates from a large pool of applicants.[35]

German psychologists had proved their military usefulness at least to the extent that after the end of World War I a small psychological testing service was carried over into the army of the Weimar Republic. That service was concerned with skill testing for special occupations. But when its head, J. B. Rieffert, seized the opportunity presented by the altered situation in 1927 to propose a reorientation to personality testing for officer candidates, his proposal was accepted with alacrity.

To appeal to their superiors, German military psychologists had to use the right language. "Character" had long been an attribute highly valued within the tradition of Prussian–German militarism. It was considered to be crucial for a successful military career. So it was "character" that psychologists had to claim to be able to assess if they were to be of use in the selection of future officers. But how were they to make good on this claim? American pencil-and-paper tests would have produced nothing but contempt and derision among their military superiors who were the survivors of total war. To be at all credible the psychological testing situations had to simulate at least some features of real military life. Accordingly, situations were invented in which the candidates were put under considerable stress, having to carry out difficult tasks while under observation, and having to function as group leaders in competitive situations.[36] Psychologists did not simply function as uninvolved observers in these situations but intervened to challenge the candidates and generally to make the tasks more demanding and stressful.[37]

These were situations then in which officer candidates might be expected to show their qualities of "character." But these qualities still had to be assessed. For this purpose the German army psychologists fell back on the only practical proposal available to them, that of the psychology of expression. Their observation of officer candidates concentrated on expressive behavior, like voice quality (Rieffert's special field), facial expression, posture, and movement. Such observations were regarded as providing direct access to characterological qualities in the manner previously discussed.

Thanks to its military development, an approach that otherwise might have remained a philosophical curiosity was transformed into a significant form of psychological investigative practice: The procedures worked out in the military context not only served purely practical purposes; they also provided the basis for German academic personality psychology during the period leading up to World War II and for many years after its end.[38]

This development was greatly accelerated by the policies of the Nazi government that assumed power at the beginning of 1933. Its frantic expansion of the German armed forces led to the creation of a large number of vacancies in the officer corps and a relatively huge increase in the number of army psychologists required for the task of officer selection.[39] The German armed forces became by far the largest employer of psychologists in the country. By 1941 the discipline achieved what had always been so difficult in Germany – namely, its official recognition as an autonomous profession through the institution of a state examination in psychology. The selection of officers for the Nazi army had finally provided the discipline with an opportunity to prove its usefulness to the centers of social power.[40] It should be noted, however, that the actual practices on which this demonstration was based were a direct continuation of the pattern that had been established during the Weimar Republic. The Nazi influence on German psychology reinforced certain already existing tendencies and put an end to others by forcing their representatives into exile.

This is not to say that there was any lack of congruence between the cultural values expressed in the psychologists' practices and those incorporated in the ruling ideology. In their day-to-day work, the German army psychologists had to make characterological attributions on the basis of the expressive behavior of the candidates that they observed. But these attributions required a network of characterological categories without which nothing could have been reported. Naturally, these categories had to be relevant to the practical purpose in view – the selection of men who were likely to carry out their tasks well in time of war. So it comes as no surprise to find that the category of "will" played a central role in these characterological systems. The way in which this category was interpreted depended strongly on the traditions of Prussian militarism.[41] "Will" was supposed to express itself in such things as a taut and upright posture, and in a particular way of shouting commands.

Why there were different psychologies of personality

The key point is that psychologists had no way of talking about human personality that was not based on cultural preconceptions *and* on the interests of particular groups within that culture. American psychologists might rate on "ascendance" and German psychologists might intuit qualities of "will," but their dependence on local cultural definitions of their task was equally profound. This dependence was not only conceptual, it extended to the specific practical purposes that served as a point of departure for their work. Their construction of "personality" or "character" as an object of knowledge was strictly confined by the rather severe limitations of the social context in which their investigations originated.

Language was always the key to the problem. Practical psychologists had to operate with verbal categories that could be used in their communication with those who provided a real or potential market for the results of their work. But that meant that the interests and preconceptions of the consumers of psychological knowledge tended to be reflected in the verbal categories used by the psychologists.

However, psychologists also needed unambiguous categories that would permit effective communication among themselves. The two sets of requirements were not easily reconciled, for the everyday terms on which the lay public insisted were incorrigibly vague and ambiguous. The American psychologists were rather more successful in meeting the requirements of a relatively unambiguous intradisciplinary language, whereas the German psychologists were perhaps more successful in appealing at least to the special section of the lay public that mattered to them. They were then left with the unresolved problem of developing a general and unambiguous language for intradisciplinary communication,[42] while their American counterparts were struggling to relate their personality-test scores back to the real world. Neither group progressed beyond an essentially descriptive network of terms to develop anything that could legitimately be called a *theory* of personality, unless by theory one simply means a set of implicit presuppositions.

In terms of their investigative practice there were of course profound differences between German and American personality psychologists. The latter kept the person under investigation at arm's length, using him or her only as a source for the real object of investigation, which was a quantitative score. The Germans, on the other hand, interacted with their candidates quite intensely over a period of several days, taking anything but a neutral observer's role. They saw their main task as that of creating a particular human situation, not that of producing a quantitative result. The qualitative nature of their methods was perfectly adapted to their social task, which was the selection of an elite that would function as the perfect representative of certain culturally defined values.[43] Selection criteria for this operation were mainly derived from Prussian traditions about what characterized a good officer; it is astounding how primitive and halfhearted were the attempts at verifying what personal qualities, if any, actually made a difference in combat situations. The methods of the German army psychologists were methods designed to select individual models of a certain kind of human being.

This had no necessary connection with morale on the battlefield, which depended on a multitude of social and situational factors. Ironically, these military psychologists were far more effective in helping to build a peacetime army than in helping to win a war.[44] In the end they had to admit that they were unable to predict courageous conduct, a failure they attributed to the difficulty of assessing "will."[45] It did not occur to them that there never was any such thing, that their basic assumption of a fixed inner character that accounted for social conduct in many particular situations was mistaken.

Psychological selection can be based on two kinds of considerations. On the one hand, there is the question of whether a selected group performs discriminably better in a criterion situation than an unselected group. This essentially statistical question can be answered without the slightest insight into the reasons for any detected difference. The considerations here are purely *technical*. On the other hand, *ideological* considerations usually enter into the selection process as well. There are all kinds of culturally sanctioned preconceptions about who is likely to succeed in what situations.

Psychological selection often took these preconceptions for granted. But beyond that there is the purely ideological selection for which the history of German personality psychology provides the most striking examples. Certain individuals may be considered worthy of selection because they represent a valued or desirable human type. If this is the kind of selection one is interested in, one's methods of choice will be qualitative. Under these circumstances, the main social function of psychological procedures will be the affirmation of the particular cultural values that have been built into them.

Although there are clearly differences in degree, no psychological assessment can altogether escape the ideological component in selection. As long as psychological performance is summarized in categories to which verbal labels are attached, an evaluative component is unavoidable. Attaching these labels to "traits" rather than to aspects of "character structure" makes no essential difference to the nature of the problem.

Lost continent: Lewin's Berlin group

The Weimar period in German history was a period of unrelieved instability and intense social conflict. German society was divided into mutually antagonistic interest groups, few of which felt any loyalty to the weak state that had emerged from the defeat of World War I. During the latter part of the Weimar period these social conflicts developed into a more and more pronounced polarization of cultural and political life. Rather unsurprisingly, this polarization recurs on the level of psychological personality study. This was a part of psychological investigation that was intensely sensitive to prevailing cultural imperatives, all the more so as it was not based on any previously existing autonomous research tradition.

We have seen how the pattern of investigative practice based on the link between characterology and the psychology of expression was involved in the temporarily profitable alliance between a section of the German psychological community and reactionary militarism. But there were other sections of this community, including all its Jewish members, for whom such an alliance was out of the question. In this group more traditional patterns of investigative practice maintained themselves, with one exception.[46] That exception was located at the Psychological Institute of the University of Berlin, where a group of young investigators, centered around the person of Kurt Lewin, developed a distinctly innovative form of practice in the field of personality research. Although the common intellectual categories of this group were based on German psychology and philosophy, their social background was more often than not female, Jewish, and eastern European. This was a rather cosmopolitan group whose spontaneous sensitivity to and awareness of social psychological questions were obviously far greater than tends to be the case in more homogeneous and more locally rooted groups. It is not surprising that their approach to the study of personality was marked by an intense sense of the embeddedness of the personality in social situations. This was a major source for the innovative elements in their investigative practice.

In the first place, the practice of this group represented a clean break with the static personality model implied in German characterology and the equally static American trait psychology. The behavior of persons in investigative situations was no longer seen as an expression of their character or as a manifestation of their personality traits, but as a part of the situation in which the person was embedded. The whole idea of a psychology of character or personality as such was inappropriate from this point of view, for the object of investigation here was not the isolated individual but the person-in-a-situation.

A different relationship between experimenters and subjects was a crucial condition for the practical constitution of this novel object of investigation. The production of abstract individuals as objects of research had been the common purpose of American trait psychology and German characterology. The former achieved this effect by reducing the interaction of investigators and subjects to a minimum and then pretending that the behavior of subjects was emitted in a social vacuum. The German representatives of expressive characterology achieved an analogous social reduction by a monadic model in which the subject's behavior was no more than an individual symptom that could only be understood as such by the investigator, but which had merely an accidental relationship to the investigative situation.

By contrast, the investigations of the Berlin group included the interaction between experimenters and subjects as part of the object of investigation from the beginning.[47] That interaction was not regarded simply as part of the technology required to evoke the response that would be the real object of investigation. Because the object of investigation for the Berlin group was not the isolated individual but the person-in-a-situation, the relationship of subjects and investigators, being part of the situation, was necessarily part of the object of investigation.

For example, in the famous study of anger by Tamara Dembo[48] the changing relationship between experimenter and subject in the course of the experiment is analyzed in some detail because it is a crucial part of the anger-producing situation in which the subject finds herself. Faced with a very difficult or insoluble problem, experimental subjects begin by asking for help, which is not forthcoming. They may then try to test the experimenter's authority, divert her attention, or show exaggerated submissiveness, until eventually an overt conflict situation becomes unavoidable. The personal affect of anger arises in this situation and obtains its meaning from it. Anger is a phenomenon that is situated in a social field formed by the power of the experimenter and the subject's efforts to come to terms with it.[49] In this type of analysis the social practice of investigation for the first time becomes an object of research interest. However, unlike much later developments, in which the "social psychology of psychological experiments" is pursued as a separate topic, there is no division between the social and the technical aspects of experimentation. The investigative situation is recognized as intrinsically social, and in research on personality it must form part of the object of investigation.

From the perspective of the Berlin group it made no sense to try to limit the interaction of experimenters and subjects to a minimum, as had seemed desirable to investigators that sought to approximate the practices of natural science in their research. Because the objects of natural scientific research cannot "answer back" or give an intentional account of themselves, many psychological investigators had sought to approximate this state of affairs in their relations with their experimental subjects. For Kurt Lewin and his group, however, this was an absurd methodological precept for the science of psychology. Even before the emergence of the Berlin research program, Lewin had rejected the arguments for minimizing experimenter–subject interaction. Taking the then current method of "systematic experimental introspection" (see chapter 3) as his point of departure, he showed himself to be intensely aware of the crucial role of communication in the methodology of introspection.[50] He rejected the incorrigibility of subjective reports and defined the aim of such studies in terms of a *mutual understanding* among experimenters and subjects – not in terms of the isolated individual consciousness. From this point of view the experimenter had to make an active contribution to the dialogue that formed the basis of the experimental results. Along the same lines, the members of Lewin's research group proceeded to emphasize the dangers in a "passive," rigidly rule-following model of experimenter behavior.[51]

The novelty in the investigative practice of Lewin's Berlin group lay in the fact that it departed fundamentally from the natural scientific model of experimentation and substituted for this a style of experimentation that attempted to do justice to the special problems of conducting empirical research on human persons.

One aspect of this turn toward a more specifically *psychological* kind of experimentation involved the rejection of the "arithmomorphic" division of human behavior that was discussed in chapter 9. There was to be no attempt to superimpose a unitizing schema on human action that would lead to countable identical elements, for the psychological significance of any action element would vary with the context:

> One ought not to regard the single trial as an isolated formation that is only identical with preceding and succeeding trials of the same kind because that is what is required for purposes of statistical treatment. Rather, one ought to treat each trial essentially nonstatistically, as a

single concrete process in its full reality. One will have to take into account the particular position of each trial in the temporal series of trials and partly make the transition to constructing [*gestalten*] the experimental period as a unitary whole.[52]

Human actions form structured wholes often extending over considerable periods of time, and any elementary component of such action sequences obtains its psychological meaning from the action structure in which it is embedded. Thus, psychological experimentation must be dedicated to the discovery of essentially qualitative relationships.

Just as the elementaristic deconstruction of human action is rejected in this type of investigative practice, so is the reduction of the environmental context to specific "stimuli." The environmental field within which a person acts forms a structured whole in which each element obtains its meaning from its relation to the entire structure.[53] An attempt at studying the investigative object person-in-a-situation cannot be based on establishing relationships between specific behavior elements and specific stimulus elements.

If the meaning of any stimulus or behavior element depends on its embeddedness in larger structures, it follows that the same action component can have an entirely different psychological significance, depending on the context. This means that measures of performance against external criteria, as favored by psychological research since the beginning of the century (see chapter 9), are at best useless and at worst misleading if one is trying to grasp the psychological significance of an action. Even in the acquisition of such an apparently straightforward skill as typing, the cumulation of successive performance measures distorts the psychological reality. For at the beginning of the process of skill acquisition, the action of typing is embedded in a psychological situation that is very different from the situation when the skill is better established, and this situation is different again from the one that exists for the fully trained typist.[54]

It is clear that a form of investigative practice dedicated to understanding human behavior in its personal and social context cannot lead to results expressed in terms of relationships among independently defined variables. Does this mean that the aim of psychological research is to be limited to the description of specific situations? This was an alternative that Lewin rejected in the most emphatic manner. He did so because he was convinced that psychologically significant situations have a depth dimension that makes all surface description inadequate. Lewin had served his apprenticeship as a psychological investigator within a research program dedicated to the experimental study of volition. That program had strong ties to the methodology of "systematic experimental introspection" (see chapter 3), a term that was in fact coined by the main representative of the program, Narziss Ach.[55] Accordingly, it was largely devoted to investigating the subjective *experience* of volition, although Ach also attempted to measure the strength of will. However, Lewin found that there was no relationship at all between these two aspects of volition. A less intensely felt intention could have a much greater effect than a more intensely felt one and vice versa.[56] A merely phenomenal description of psychological events was therefore inadequate.[57]

Working out the full implications of this insight depended on Lewin's very intensive studies in the philosophy of science.[58] These convinced him of the necessity of making a general distinction between the surface pattern of events and an underlying causally effective reality. He referred to this as the distinction between "phenotype" and "genotype," and it became the indispensable centerpiece of his approach.[59] Studying psychological events in context meant more than taking into account the temporal dimension of action structures. It meant penetrating beyond the apparent meaning of an action to an underlying context that was causally effective. For example, two actions might be indistinguishable on the surface, but genotypically the one might represent the carrying out of an order and the other the carrying out of a voluntary intention. The aim of psychological investigation was the exploration of the genotypical context of human actions. "Personality" or "character" as reified static entities are replaced by types of psychological context as the real objects of psychological investigation.

"Types of psychological context," however, cannot be discovered inductively.[60] Genotypical contexts are decidedly not empirical generalizations from a number of similar cases. They are theoretical constructions whose empirical verification could in principle be achieved by the demonstration of a single appropriate case. Lewin makes a distinction between "being-thus" *(Sosein)* and "being-there" *(Dasein)*.[61] The first concept refers to the nature of a type, the second to the realization of a type in empirical reality. Rationally, the exploration of the former should precede the establishment of the latter, for if one started with the collection of empirical facts, one would merely be aggregating instances on the basis of superficial resemblance. Lewin strongly criticized the widespread psychological research practice of forming statistical aggregates on the basis of "historical–geographic" criteria,[62] such as membership of specific social groups (see chapter 5). Such criteria were not psychological criteria at all, and their use to form statistical aggregates could only result in the formation of psychologically heterogeneous entities from which no psychological insights could be expected.

For the Berlin group the aim of psychological research was not the search for statistical regularities defining the relationship among isolated variables. Testing the statistical null hypothesis would constitute the polar opposite to their style of research practice. The lawfulness of psychological processes was assumed a priori; it was not something to be established by a statistical test. But the conception of lawfulness involved here had nothing to do with the statistical frequency of association among specific empirical measures. Rather, it involved the belief in the existence of specific genotypically effective "process types" *(Geschehenstypus)*.[63] These were qualitatively distinct and temporally extended patterns of events that were characteristic of specific human situations. The concrete research practice of the Berlin group was dedicated to the experimental simulation of such situations – for example, anger-producing situations or situations of psychological satiation.[64] These investigations were "experimental," not only in the sense that they represented an artificial re-creation of common human situations, but also because they included planned variations in these situations for the purpose of exploring their genotypical structure.

The Berlin group fell victim to the full destructive force of Nazism. Its members were scattered and had no chance of maintaining the extraordinary momentum that the group had developed during the relatively brief period of its existence. Lewin himself emigrated to America, where he faced several years of marginalization by the psychological establishment. His ability to address socially interesting topics brought him considerable fame during the World War II period, but he died early in 1946. As far as the distinctive aspects of his form of investigative practice are concerned, his influence on American personality research was slight. Most of the early methodologically important papers remained untranslated, and those of his students who showed any inclination to incorporate his methodological precepts into their own work were excluded from the experimental mainstream of American psychology.[65] In addition, Lewin's own preoccupation with problems of formalization during his post-Berlin period tended to divert attention from the most innovative aspects of his contribution to its more sterile aspects. From a methodological point of view, the "Lewin legacy" remained something of a buried treasure.

11

The social construction of
psychological knowledge

The preceding analysis of the historical development of investigative practices in psychology is obviously based on a certain model of the investigative process. Although various aspects of this model have emerged in previous chapters, some of the basic features of the model still need to be made explicit. The first part of this chapter is devoted to this task. In the remainder of the chapter I briefly address some general questions pertaining to the status of psychological knowledge that arise out of the kind of analysis presented here. These questions all revolve around the issue of how the social construction of psychological knowledge affects the reach of the knowledge product. Although knowledge claims are generated in specific sociohistorical situations, it does not follow that they have no significance or validity beyond those situations. There are two possibilities here. First, psychological knowledge may be generalizable to situations other than those in which it was generated. This is the question of applicability. More fundamentally, however, there is the possibility that psychological knowledge claims may be able to tap a level of "psychological reality" that is independent of the special conditions under which it is investigated. The second and third sections of this chapter discuss these possibilities.

The political economy of knowledge production

Investigative practices are part of a productive work process. Their employment results in the generation of valued products, though of course these are not primarily material but symbolic products – namely, scientific knowledge claims. This same employment necessarily involves people who put these practices to work and also some kind of raw material that is productively transformed into the valued knowledge product. Producers of scientific knowledge never work as independent individuals but are

179

enmeshed in a network of social relationships.[1] What they initially produce is not so much knowledge as knowledge claims. Such claims are only transformed into knowledge by an acceptance process that involves a number of individuals – such as reviewers, readers, textbook writers – who share certain norms and interests.[2] The anticipation of this acceptance process affects the production of knowledge from the beginning.

Like all human productive activities the production of scientific knowledge is highly goal-directed. This directive component manifests itself not only in the fact that so much scientific activity is explicitly devoted to hypothesis testing, but also in the more implicit commitment to the search for a certain type of knowledge. The investigator is not just interested in any kind of answer to specific questions; he or she is only interested in obtaining such answers on certain terms that have been set in advance. Knowledge products must take a particular form: They may have to be quantitative, for example; they may be limited to intersubjectively observable phenomena; they may be required to divide the world into precisely defined variables. These formal prescriptions, which dictate in advance what form the knowledge product shall take, often become quite specific. For example, they may dictate that only linear and additive relations among variables shall appear in the product, or that the product shall take the form of interindividual differences. As we have seen in this volume, the kind of knowledge product desired has a decisive influence on the choice of investigative procedures.

Prescriptions pertaining to the form of knowledge products are themselves social products. They are characteristic of particular groups of knowledge producers and unite their members in common methodological commitments. More fundamentally, they tend to be products of the interface between the group of knowledge producers and centers of social power that are able to determine the success or failure of their enterprise.[3] Thus, the investigators gathered in the seventeenth-century Royal Society decided to pursue a form of knowledge that promised technical utility and to exclude forms of knowledge that were tied up with questions of social reform and revolution. The end result was the modern conception of the "value neutrality" of scientific knowledge.[4] But the decision to limit the forms of admissible knowledge was made because producers of knowledge lived and worked in a social context where certain types of knowledge were perceived as useful and others as threatening to powerful social interests.

In other words, rules about the admissibility and desirability of particular kinds of knowledge product are generated in the course of a *historical* process of negotiation that involves the coming to terms of knowledge producers with the realities of social power and influence in the world they inhabit. Direct relationships with the most important centers of political and economic power are sometimes involved, although in complex industrial societies this power is increasingly mediated by secondary groups that exercise power in limited areas. Any new group of knowledge producers, like scientific psychologists, will have to come to terms with a world that already contains knowledge producers that make related claims, such as medical professionals, educationists, and experimental physiologists. These established groups wield a certain amount of social power, and that power is based on their institutionalized monopoly over certain types of knowledge product. How can a new group of knowledge producers thrive under these conditions?

It can thrive only if it manages to form effective alliances. This it can do by enlisting the interest of established groups in its knowledge products and avoiding their censure, a process that has many facets.[5] First of all, the new knowledge products had better reflect well-established preconceptions about the forms of valuable knowledge. If quantitative knowledge is particularly valued, then it helps to establish new claims if one can give them a quantitative form. But established preconceptions about the form of knowledge products automatically extend to the methods used to generate them, for the form of the product depends directly on the nature of the method of production. So one must be seen to be engaging in practices that produce the right type of knowledge, even though such rituals have more in common with magic than with science.[6] More specifically, the borrowing of investigative practices from better-established fields, like experimental physiology, provides a basis for limited alliances that link the new field and its products to the existing network of recognized scientific knowledge.

But this kind of alliance would provide a very limited basis for autonomous development if the new knowledge products were not seen as having a significant social value of their own. They must become marketable, and that means that there must be categories of persons to whose interests the new product is able to appeal. The more powerful and better organized these consumers of knowledge products are, the more successful the producers will be in consolidating their own position. American psychologists scored some real successes in this direction by providing knowledge products that mobilized the interests of educational and military administrators as well as the administrators of private foundations.[7] Even if the alliances so formed were often temporary, and based more on promissory notes than on real goods, they served an important function in establishing the credentials of the new discipline at a critical stage in its development.[8]

The relatively new knowledge products that scientific psychology had to offer were hardly dumped on an empty landscape but had to be carefully inserted into a terrain whose commanding positions were already occupied. There were important national differences in regard to who occupied the commanding positions and how firm the occupation was, as we saw in chapter 8. But the crucial point is that the successful establishment of a new discipline is very much a *political* process in which alliances have to be formed, competitors have to be defeated, programs have to be formulated, recruits have to be won, power bases have to be captured, organizations have to be formed, and so on. These political exigencies necessarily leave their mark on the discipline itself, and not least on its investigative practices. The political environment largely determines what types of knowledge product can be successfully marketed at a particular time and place. The goal of producing an appropriate type of product plays a crucial role in the selection of investigative practices within the discipline. Ultimately, the limitation of investigative practices results in corresponding limitations on the knowledge products that the discipline has to offer. So in the end, those with sufficient social power to have an input into this process are likely to get the kinds of knowledge products that are compatible with their interests.

Being obliged to operate in an external political environment, the community of knowledge producers is bound to develop an internal political environment. In the face of the external world, this community constitutes a special-interest group, but these special interests presuppose a more general system of knowledge production within which they are recognizable as special interests. There is an intimate relationship between the general forms of presuppositions, knowledge goals, and investigative practices and their specific embodiment.[9] As the community of knowledge producers grows it develops internal norms and values that reflect its external alliances. Its professional project is directed at carving out and filling a particular set of niches in the professional ecosystem of its society, and its internal norms reflect the conditions for the success of this project.[10] These norms tend to govern both the production of knowledge and the production of the producers of knowledge through appropriate training programs.

Individual research workers tend to take these norms for granted and therefore believe they are making rational and technical decisions about the methodology they adopt. They forget that the historical development of the discipline has preselected the kinds of alternatives realistically available to them. In the case of psychology, it is particularly implausible to maintain that this preselection was rationally determined. Given their historical situation, individual investigators may well be making rationally defensible choices about their research practice, but there is no guarantee that the sum of these choices somehow constitutes a rational development of the discipline as a whole. The history of twentieth-century psychology provides little comfort for believers in the existence of a rational "hidden hand" that steers development on the disciplinary level. If one accepts sharp limitations of time and place, and if one stays within a very restricted framework of assumptions, one can make a case for some spurts of more or less rational development. But if one adopts a broader perspective, the political fundamentals become impossible to ignore.

This is also the case when one examines the investigative situations set up in the course of disciplinary practice. These have a political dimension because they necessarily involve questions of rights and questions of power. In the type of investigative practice that came to dominate the discipline, investigators are clearly in a position of power and authority vis-à-vis their human subjects. They give the directions, they manipulate the situation, they control the setting, it is their purposes that determine the course of events, and so on. But there are limits to what they can get away with, because ultimately the human and civil rights of experimental subjects must be recognized if unpleasant political consequences for the profession and legal consequences for its individual members are to be avoided. In the language of professional ideology, the rights of subjects come under the heading of research ethics while the powers of investigators come under the heading of technique. But of course both are part of the political aspect of the research process and are closely related.[11]

Investigative practices involving human subjects have always involved implicit political decisions (see chapter 4). Once a certain investigative style had established its hegemony, the question of subjects' rights became largely restricted to negative freedoms, such as limits on what could be done to subjects or limits on permissible degrees of ignorance about what was going on. But, as we have seen, in an earlier period the positive rights of subjects were an important issue. In particular, there was the question of whether the experience of subjects was to count within the framework of the investigative situation. The Wundtian experiment, the psychoanalytic situation, the Gestalt and Lewinian experiments all answered this question in the affirmative. However, it became a key principle of the dominant model of psychological experimentation that the subject's experience was to be discounted.[12] Attempts to change this state of affairs have always evolved the most determined resistance. At one time behaviorist dogma served a purpose in justifying what amounted to the disenfranchisement of experimental subjects. When this dogma became discredited, other reasons were found. The resistance is not surprising because taking subjects' experiences seriously strikes at the basic political structure of the dominant form of psychological experimentation.

Political questions regarding participants in investigative situations are not a peculiarity of psychology. From the beginning of modern scientific experimentation in the seventeenth century, the question had to be faced of how agreement was to be reached on basic facts. The solution that won the day involved reliance on the testimony of reliable witnesses about what they had observed, and the reliability of witnesses depended on their social status.[13] This was clearly a political decision, which confirmed that the pursuit of science was to be limited to an elite. Psychologists had to face a similar decision, although in their case the consequences were probably more profound. One could do as the physical sciences had done and limit

participation in experiments to a community of elite observers whose experiences provided the basic information on which the science relied. This
pattern was followed in the Leipzig experiments and by various successors.
Alternatively, one could admit anyone to participation in psychological
investigation and then establish a kind of class system within the investigative situation, so that only the investigators counted as competent
witnesses while their subjects were limited to the role of objects of investigation. This alternative, as we know, proved to be far more successful
than the first.

The most important point, however, is that the political forms imposed
on the knowledge-generating situations profoundly affected the kind of
knowledge that was produced. Whereas in the physical sciences the social
arrangements among experimental participants did not in themselves alter
the material to be investigated, this was hardly the case in psychology.
Here, the object of investigation was itself part of the social context of
investigation. The persons, whose "responses" formed the basis for the
psychological knowledge product, were themselves part of the social context of investigation. This entailed a rather intimate link between knowledge product and investigative situation, a link that could pose a threat to
the generalizability of psychological knowledge. To what extent could such
knowledge be applied in social situations that differed from those in which
it was produced? This question requires some consideration within the
present framework.

Expert knowledge and life outside the laboratory

The relationship of the producers of scientific psychological knowledge to
other knowledge producers is not only a question of interprofessional alliance and competition. Before such questions can become significant in
the development of a field, it must achieve a minimal status as a science.
It must have earned sufficient respect as a knowledge-producing enterprise
to be a serious candidate for competition or for alliance with other pro-
fessionalized fields. That means its products must have become clearly
distinguished from the everyday or common knowledge and belief of the
lay public and achieved the status of *expert* knowledge.[14] There is nothing
more inimical to a field's success as a source of valued knowledge than the
suspicion that it is able to supply no more than a duplication of what
"everyone" knows anyway or, worse, a reinforcement of popular super-
stitions. The "soft" areas of psychology and parapsychology have always
had to contend with this problem.

There had to be constant vigilance about maintaining the sharpest possible differentiation from folk knowledge. This was not a problem limited
to some marginal areas but affected the discipline as a whole. In the judgment of the lay public everyone had to rely on psychological knowledge

in making his or her way through the world. How could anything offered by experts compete with a lifetime of experience in human affairs? This ever-present background challenge had effects on the investigative practices of psychologists that should not be underestimated. Whatever else they may have done, these practices also served to demonstrate a crucial *distance* from those mundane situations in which everyday psychological knowledge was acquired. This was achieved largely by drawing on the mystique of the laboratory and the mystique of numbers, both of which had been well established prior to the appearance of modern psychology. The very artificiality of laboratory situations became a plus in establishing the credentials of knowledge claims emanating from this source, and the imposition of a numerical form on otherwise trivial knowledge gave it an apparent significance with which lay knowledge could not compete. Replacing ordinary language with jargon helped too.[15]

Nevertheless, the relationship between expert and everyday knowledge had other aspects. For one thing, investigative practices involving human subjects were unavoidably social practices and as such were not as clearly distinct from extralaboratory practices as the experimental mystique made out. In order to generate psychological knowledge, if one was not to rely solely on self-observation or on animal analogies, one had to set up social situations involving clearly patterned relationships among the human participants. These situations resembled certain more familiar social situations, not only because they were born in already established institutional environments, but also because the type of knowledge product desired had affinities with other knowledge products gathered in those environments. As we saw in chapter 4, clinical experiments emerged in a clinical environment that had preformed the social relationships among investigators and their "subjects" as well as the shape of the knowledge product that resulted from their interaction. Similarly, mental testing mimicked the social situation of a school examination in a recognizable way, and its product, the objective grading of individual performances, represented the optimal result of an idealized examination. In the last analysis psychological investigative situations constituted a development of already existing social practices.[16]

This was just as well, for the relevance of psychological knowledge products to life outside the research setting depended on this link. Clinical knowledge was applicable in clinical situations, and the knowledge gained by means of mental tests was applicable in settings devoted to the grading and selection of individuals on strict performance criteria. In these cases the homology of the investigative and the application setting made possible a direct transfer of information from the one to the other. Where this homology did not exist, the general relevance of research products required an elaborate apparatus of speculation, which made the whole enterprise an easy target for criticism. Often the viability of generalizations from research depended on the effectiveness of certain ideological presuppositions. Thus, whereas the application of results from intelligence tests to selection problems in schools was no more problematic than the use of any other kind of examination result, more general inferences about human intelligence were bound to be far more controversial. They depended on elaborate assumptions about the structure of the human mind and about the role of heredity in human affairs. Generalizations from research in other areas also depended on the plausibility of certain fundamental assumptions.[17]

One assumption was so pervasive in generalizations from psychological research that it deserves special mention. The nature of this assumption is revealed by the way in which the regularities emerging out of investigative situations were reattributed. They were attributed, of course, to the individuals who participated in the investigation as subjects. However, the regularities in question were not taken as conveying information about an individual-in-a-situation but about an individual in isolation whose characteristics existed independently of any social involvement. This Robinson Crusoe assumption played an increasingly important role the more psychology turned from psychophysiological investigations to generalizations about human conduct on the molar level. Faith in the Robinson Crusoe myth made it seem eminently reasonable to ignore the settings that had produced the human behavior to be studied and to reattribute it as a property of individuals-in-isolation.

Psychologists were not the inventors of this myth.[18] Their discipline was a relatively late product of an individualistic civilization that had long cherished the ideal of the independent individual for whose encapsulated qualities all social relations were external. In adopting this myth as a fundamental and unquestioned assumption for their work, psychologists were showing themselves to be no more than good citizens of the society that nourished them. Yet their role was not entirely passive. For by taking the myth for granted in a context that was supposed to be devoted to objective knowledge, they were in effect giving it a new justification and legitimacy. Their procedures for interpreting the products of their investigations were a kind of enactment of the founding myth that they shared with their compatriots. Moreover, psychological research often incorporated individualistic presuppositions in a uniquely unadulterated and often crass manner. More than the contributions of other social scientists, the work of psychologists represented a kind of celebration of the myth of the independent individual in its pure form.[19]

Scientific psychology found itself in this position, not only because of the nature of the subject matter that had been bequeathed to it from its prescientific phase, but also because the natural scientific methodology, on which it staked its prestige, converged nicely with its traditional assumptions. Experimental methods isolated individuals from the social context of their existence and sought to establish timeless laws of individual behavior by analogy with the laws of natural science. Shared social meanings and relations were automatically broken up into the properties of separate individuals and features of an environment that was external to each of them. The functional relationships established in this way presupposed an individualistic model of human existence.[20] Anything social became a matter of external influence that did not affect the identity of the individual under study.

This division of the actions of socially situated persons into separate individual and external situational components provided the framework for a general atomization of the features of both the individual and the situation. The latter had become an "environment" for the independent individual. A pursuit of universally valid natural scientific laws most often took the form of establishing empirical associations between separate individual and environmental "variables" and "factors." This enterprise depended on the crucial assumption that the variables studied would retain their identity irrespective of context.[21] Discounting of context applied both on the individual and on the situational level. Only by assuming that the responses of several individuals to a psychological instrument were manifestations of an identical psychological property that remained qualitatively the same, irrespective of differing personal contexts, did it make sense to combine individual responses into the kind of abstract collectivity that was discussed in chapter 5. Similarly, it only made sense to develop theories about psychological features like learning in the abstract, anxiety in the abstract, or intelligence in the abstract, if one assumed that such features remained identical irrespective of the situations in which they were investigated.[22]

Applications of psychological research products tended to fall along a scale from specific applications in mundane situations to grand claims about human nature that had profound implications for social policy. Among the latter one would find claims about the existence and heritability of "general intelligence" or claims about the origins and social role of race prejudice. At the other end of the scale there would be local decisions about individuals on the basis of test results, or improvements in human engineering. Now, it is easy to see that the two kinds of application differ in terms of the distance between the investigative situation and the situation to which the research products are to be applied. In the case of grand applications to social policy there is a vast distance between the situations in which the research products are generated and the kinds of situation to which they are to be applied. This distance has to be bridged by a host of unproven and often unspoken assumptions, and thus the whole enterprise is essentially based on leaps of faith. But in the case of narrow, technical applications of psychological knowledge there is typically a considerable degree of continuity between the investigative situation and the situation in which the application takes place. When one applies intelligence- or aptitude-test results to the prediction of future performance in the appropriate settings, academic or otherwise, one is essentially using a *simulation* technique. The more effectively the investigative context simulates the context of application, the better the prediction will be.[23]

During the period under review in this volume such a formulation would not have been acceptable, or perhaps even comprehensible, to most psychologists because of their metaphysical commitment to the reality of a world made up of fundamentally independent individuals. With such a starting point there were two kinds of approach that psychologists could adopt toward their subject matter. They could either set about studying concrete individuals and their specific attributes, or they could study abstract relationships that held universally across individuals. The former alternative was incompatible with conventional notions of scientificity, so the second alternative carried the day. Eventually, this long-recognized alternative was named and codified in terms of the well-known distinction between the idiographic and nomothetic approach.[24] As understood by most psychologists, this distinction generally served to rationalize their search for universal "laws" or principles. This was because the division of methods into idiographic and nomothetic was presented as though it exhausted the possibilities, and because the idiographic alternative was given a strictly individualistic interpretation. Faced with the choice between the study of individual biography and the study of "scientific laws of human behavior," most psychologists had no difficulty in deciding on the more ambitious alternative.

Unfortunately, this set of false alternatives removed the most promising possibilities from view. Those possibilities open up with a recognition of the fact that a series of situations, including investigative contexts, can have structural features in common. Individual behavior, being always part of such situations, would then show common features across a range of structurally similar situations. This would make generalization and prediction from one context to another possible, although such generalizations would always have to be regarded as conditional on the comparability of situations in terms of the similarity of their crucial structural features.

However, we should note that there are two ways in which the distance between the context of investigation and the context of application can be narrowed. Letting the context of investigation simulate the context of application is one way, and certainly the more easily recognized way. But in many cases of successful application of psychological knowledge one can also detect a reciprocal process that involves a change in the context of application so that it comes to resemble the context of investigation. The conventional model of the application of scientific knowledge involves a fixed context of application, but of course this is not true in the real world. Rather, the application involves an actual change of practices, which are transferred from the artificial laboratory setting to the world outside. In other words, the application of knowledge is possible only insofar as an artificial construction, derived from the investigative situation, is imposed on the natural world. This is what happens in the construction of industrial plants or in the adoption of sterile conditions in medicine, for example.[25]

An analogous process operates in the case of psychological knowledge. Applying mental tests often meant modifying some of the practices of schools so that they resembled testing practices more closely. Having spawned intelligence tests in the first place, some school examinations subsequently imitated their offspring in terms of such matters as timing. question format, and use of statistical norms. Or, to take another example from a later period, the adoption of behavior-modification programs in institutions involved the restructuring of the context of application so that it more closely resembled a particular investigative context. Practices have always wandered from noninvestigative to investigative contexts and back again. Genuine application of psychological knowledge depends on this, for the bond between knowledge and the practices with which it is associated is an extremely intimate one. Abstract knowledge only exists abstractly; its application requires a transfer of corresponding practices. So if the purely ideological application of psychological knowledge in support of particular social policies were ever to be converted into a real application that affected its objects directly, it would require an appropriate reconstruction of society, as B. F. Skinner so clearly demonstrated in *Walden Two*.[26]

It is apparent that the question of the distance that separated investigative situations from the situations in which their knowledge products were applied involved psychologists in a serious contradiction. On the one hand, the rhetoric of science required that this distance be emphasized and magnified. Because the yield from investigative situations was supposed to consist of universally valid generalizations (so-called nomothetic laws), these situations were endowed with a mystique that rendered them so remote from ordinary life that they were not even seen as social situations. Even the idea that there might be a social psychology of psychological experiments only arose at a late stage in the development of the discipline.[27] However, there remained the rather indigestible fact that the discipline's ability to make fairly reliable predictions about human beings outside the psychological laboratory depended to a large extent on the closeness of the context of investigation to the context of application. It was often the case that psychological knowledge had some technical utility only insofar as its investigative practices were continuous with relevant social practices outside the investigative situation.

This contradiction between scientistic rhetoric and the facts of life in applied psychology tended to maintain the separation of "pure" and applied research. Politically, both the rhetoric of science and specific technical utility were indispensable for the rapid development of the discipline and both continued to flourish side by side. But the political effect was enhanced by a professional ideology that interpreted any successes of so-called applied research in terms of the "application" of knowledge derived from "pure" research.[28] In this way each partner could profit from the value attributed to the activities of the other (see chapter 8). "Pure" research could claim support on the basis of its ultimate practical usefulness and "applied" research could speak more authoritatively by clothing itself in the mantle of science. In actual fact, "applied" research usually relied on its own practices with little or no help from "pure" research, and "pure" research either continued in complete isolation or adapted the practices of "applied" research, contributing little but technical sophistication and a more abstract terminology. Throughout the period under review here, the major function of laboratory research was probably the professional socialization of aspirant members of the community of producers of psychological knowledge. Laboratory work imparted a very specific interpretation of the meaning of "science," and one must suspect that most of the products of this work existed mainly to fill the pages of textbooks.

Below the rhetoric of pure and applied science it is possible to discern a developing fundamental convergence between contexts of investigation and contexts of application. In the investigative contexts that became increasingly popular during the first half of the twentieth century, the individuals under investigation became the objects for the exercise of a certain kind of social power. This was not a personal, let alone violent, kind of power, but the kind of impersonal power that Foucault has characterized as being based on "discipline."[29] It is the kind of power that is involved in the management of persons through the subjection of individual action to an imposed analytic framework and cumulative measures of performance. The quantitative comparison and evaluation of these evoked individual performances then leads to an ordering of individuals under statistical norms. Such procedures are at the same time techniques for disciplining individuals and the basis of methods for producing a certain kind of knowledge. As disciplinary techniques the relevant practices had arisen during the historical transformation of certain social institutions, like schools, hospitals, military institutions, and, one may add, industrial and commercial institutions.[30] Eventually, the knowledge-generating potential of these kinds of practices became realized in an increasingly systematic way, and the knowledge so produced was fed back into the original disciplinary institutions to increase their efficiency. This kind of knowledge was essentially administratively useful knowledge required to rationalize techniques of social control in certain institutional contexts. Insofar as it had become devoted to the production of such knowledge, mid-twentieth-century psychology had been transformed into an administrative science.

The limits of constructivism

The profound dependence of the psychological knowledge product on the nature of the practices that generated it did not hurt the "external validity" of the product.[31] As we have suggested, that depended on the similarity of the context of investigation and the context of application. Because the common practices of investigation had arisen by an extension of previously existing social practices there were always certain areas of life that had already been prestructured for the insertion of a certain kind of psychological knowledge. The fact that psychological knowledge was constructed, rather than discovered by naive inspection, did not mean that it had no general validity. As long as the principles used in the construction of the knowledge product were not totally unique, the possibility of at least a degree of generalization beyond the investigative situation was always present. However, this possibility was always circumscribed by the extension of the relevant social conditions.

Psychology of course aimed at a different kind of generality for its knowledge products. Large parts of it, at any rate, pretended to a knowledge that was universal, ahistorical, and independent of both the context of investigation and the context of application. Such knowledge almost always turned out to be chimerical, although even in this form it may have had an important social function. It may have provided a pseudoscientific affirmation of various ideological positions that had been presupposed in the construction of the knowledge in question, particularly an ideology of abstract individualism and an ideology of social control.[32] On this level generalizations from psychological research also referred to a certain reality outside the research situation. Although it was only the reality of an image of human nature and society, faith in such images could have material social consequences. In the long run, the contribution made to this kind of faith may have represented a more successful application of psychological knowledge than any technical application in specific institutional contexts.[33]

Neither the technical nor the ideological validity of psychological knowledge provided it with more than local significance. Technically, its relevance depended on the existence of a certain institutional framework, and ideologically its plausibility extended only as far as the cultural forms in which the shared faith was expressed. The question that remains is whether the investigative practices of psychology could ever yield a kind of knowledge whose significance was more than local. In this volume we have been examining the dependence of the knowledge product on the conditions of its production, and this has necessarily entailed a deconstruction of the generally false claims to universality that were commonly made on behalf of psychological knowledge. Does this mean that such claims were always and necessarily false? When allowance is made for the factors that led to a relativizing of psychological knowledge, is there no remainder?

Let us be clear that this question is not concerned with knowledge about the physiological conditions of human action and experience but with specifically psychological knowledge. A relatively small proportion of research within the discipline of psychology has always taken these physiological conditions and their effects as its subject matter. But psychology became an autonomous discipline only when such investigations were banished to a corner of the field and attention was concentrated on purely psychological knowledge. The question is whether the latter ever managed to have reference to a reality that was not simply a product of the investigative context.

Such a reality would have to exist on some level other than that of the regularities established within specific investigative situations. For these regularities, we have seen, are largely a product of various investigative practices. If the history of psychological research demonstrates anything, it demonstrates the extraordinary pliability of human beings. Of course, there are limits to this pliability. There are obvious physiological limits, though these are not particularly problematical. Psychological limits that are themselves a product of the investigative context, as when an experimental task is too fatiguing or boring or confusing, can easily be accommodated by a constructivist paradigm, as can those limits that are a function of the particular social background of experimental subjects. But are there general *psychological* sources for the limitations on human pliability in investigative situations? Such sources would constitute a level of psychological reality that is not to be confused with the overt regularities observed in the course of investigation.

As long as psychological reality is identified with the overt regularities of behavior established in investigative situations, there can be no escape from the consequences of a purely constructivist perspective. Such regularities are undoubtedly a product of social construction. However, constructivism may not have the last word if we recognize the fundamental distinction between what, in Bhaskar's realist philosophy of science, is called the domain of the real and the domain of the actual.[34] The latter comprises the observable events that, under specific conditions, can form the basis for empirical domains. The domain of the real, however, involves generative mechanisms that are not observable in themselves and that exist independently of any investigative intervention. Scientific research, whether in psychology or in physics, aims to increase our knowledge of such generative mechanisms.[35] Observed regularities are only of interest insofar as they assist in this task.

In the earlier phases of the history of modern psychology this kind of perspective was not uncommon. Wundt himself introduced the concept of *psychic causality,* which covered the psychological generative principles that were at work both in psychological experiments and in everyday life.[36]

Psychic causality made the difference between organismic responding and the construction of an organized subjective world. Although Wundt declared the exploration of psychic causality to be the aim of psychological research, there was a huge gap between such aims and his methods, so that his conception of psychic causality remained largely a matter of good intentions. Nevertheless, he had boldly confronted a fundamental issue which most of his successors preferred to sidestep.

Another psychologist who recognized the crucial importance of psychic causality, or psychic determinism as he called it, was Freud. He consistently tried to explain his clinical observations, as well as certain patterns taken from "everyday life," in terms of an underlying level of psychological causal processes. His version of psychological causality was very different in content from that of Wundt, but it was also far more specific. Nevertheless, the relationship between the level of empirical observation and the level of generative mechanisms remained problematic.

Gestalt psychology and the related Lewinian psychology also depended on notions of psychological causality that now took the form of models for the operation of forces in psychological fields.[37] These models were always underdetermined by the empirical evidence; they referred to a domain of causes that was different from the domain of observations. Toward the end of our period Jean Piaget began to develop yet another approach to the problem of psychological causality but his model did not become widely known until much later.

These and other less prominent attempts at coming to grips with the issue of psychological causality foundered on two interdependent problems. The various conceptions of the nature of psychological causality were clearly widely divergent and their ties to their respective empirical domains were insecure. Neither of these problems was peculiar to psychology. Fundamental differences about the nature of physical causality were far from being an unknown phenomenon in the history of physics, and even the underdetermination of theoretical models by empirical domains had never in itself been a cause for panic. However, in the historical context of a discipline desperate to establish its status as a reliable source of certain and practically useful knowledge, these fundamental dissensions were easily seen as threats. The ascendancy of positivist philosophies of science during the critical formative period of the discipline also meant that its struggle for legitimacy had to be fought in an intellectual environment that was hostile to conceptions of reality that went beyond the phenomenal level.[38]

It became common to identify psychological reality with empirical regularities. The gradual adoption of the rule that all theoretical concepts had to be "operationally defined," and the peculiarly narrow interpretation of this rule, guaranteed that the very possibility of a domain of psychological reality beyond the domain of empirical regularities would never be considered. But there was something supremely ironical about this outcome. For in its flight from a domain of reality stigmatized as "metaphysical," the discipline had limited itself to a kind of reality that turned out to be peculiarly vulnerable to the corrosive effect of social constructivism. The kinds of empirical regularity that were the product of psychological investigation owed their existence to the investigative practices of psychologists. If these regularities also reflected a domain of reality that was independent of these practices, such a domain would have to be defined in terms clearly distinct from the constructed empirical level.

Not that models of psychic causality were anything else but the products of a constructive process, a theoretical rather than a practical construction. However, their existence makes possible the confrontation and reciprocal criticism of theoretically constructed and practically constructed domains. This confrontation offers a possibility of escape from the closed world of unreflected investigative practices but one that can be realized only if the process of criticism really becomes mutual. If traditional investigative practices are considered inviolate, so that only the one-sided criticism of theoretical constructions by the products of practical construction is regarded as legitimate, no progress toward a more reality-adequate kind of knowledge is to be expected.[39] The adequacy of investigative practices must also be measured against the requirements of divergent models of psychological reality.[40] Modern psychology arose in part out of a justified criticism of contrived speculations that were never measured against any systematically constituted empirical domain. Unfortunately, this legacy helped to promote a certain blindness toward an equally unsatisfactory possibility, namely, a discipline based on faith in a limited set of procedures whose fundamental structure is never seriously tested against alternative conceptions of psychological reality.[41]

For most of the twentieth century two circumstances made it difficult for psychological investigation to escape from the web of its own constructions. First, the "cult of empiricism" and its Humean view of reality exerted an iron grip on the discipline. Because empirical domains were necessarily special constructions, theoretical concepts that remained completely tied to them provided no access to anything but a constructed reality. Second, mainstream psychology was isolated from potentially liberating influences. Within the discipline alternative conceptions of theory, practice, and their interrelationship were successfully marginalized. At the same time, the boundaries of the discipline were jealously guarded, thus minimizing the possible impact of alternative models of psychological reality and scientific practice emanating from other social and human sciences and from philosophy. Far from opening themselves to such influences, American psychologists were able to exploit the special congruence of their founding myth with the "American ideal" to assert imperialistic claims over many other areas of human knowledge.[42] This approach did not promote critical reflection about the basis for such claims and allowed the question of access to psychological reality to be decided on a purely technical level.

If, however, we think of reality as a domain that exists independently of empirical constructions, then the question of access to such a domain cannot be decided on the level of a particular empirical investigation. That would imply an epistemic individualism according to which knowledge is the product of an interaction between an individual investigator and nature. But epistemic access to the world is collective[43] – it is always mediated by the social conditions under which groups of investigators work. Moreover, in psychology these conditions are relevant not just for the investigators but also for those who function as the objects of investigation. Like all social work, this collective enterprise is governed by definite rules that are reflected in the form of the product. The knowledge product never consists of a collection of independent elements but always takes the form of an ordered array of some kind. The nature of such an array depends on the ordering principles that are incorporated in the constituting practices.[44]

Moreover, it will have become clear that investigative practices do much more than order observations of a world that is given – they actually prepare the world that is there to be observed. The observations that are available for ordering may function as raw material within the limited framework of a particular investigation, but they are far from raw in terms of any broader perspective. They are made on human subjects who have been actively selected on certain social criteria, who have entered into particular social relations with the investigators, and whose actions have been carefully circumscribed by the investigative situation. Ordering observations once they have been made is only a small part of investigative practice; the major part involves the construction of special "forms of life" in which the observations are grounded.[45] In other words, the work of constituting knowledge domains does not take place only on the level of cognition but involves significant construction on the level of human action and social relations. Knowledge products and the experimental forms of life that generate them are intimately linked. Particular experimental actions and relationships only exist in order to produce knowledge of a certain type, and that knowledge cannot escape the imprint of the forms of life in which it originated.

All this does not mean that we should now replace the naive naturalism of the past with a simpleminded sociological reductionism. To say that psychological knowledge bears the mark of the social conditions under which it was produced is not the same as saying that it is *nothing but* a reflection of these conditions. However, we certainly will not discover what more it might be unless we take the social embeddedness of the knowledge product fully into account when raising epistemological questions. The fundamental mistake that a social constructionist approach can help to correct is embodied in the pervasive assumption that the categories we use in the course of empirical investigation correspond directly to the "natural kinds" that exist in a real world outside the framework of our investigative and intellectual practices.[46] But it does not follow from this that one set of categories and practices is as good as another – that "anything goes." There are defensible criteria for assessing the cognitive yield from the employment of any set of would-be scientific categories and practices. For example, we can make justifiable comparisons among alternative explanatory schemes in terms of their attribute of "depth," and we can assess the social and moral consequences of different kinds of knowledge claims.[47] So far from undermining such comparisons, insights into the historical and socially constructed character of knowledge products are a necessary preparation for making those comparisons.

Of course, in the day-to-day business of technical research such questions do not need to be raised. On this level *specific* knowledge claims can be compared with each other as long as one takes for granted the vast assembly of hidden practical and theoretical assumptions that they share. In other words, the propositional *yield* from the knowledge producing process can be separated from the process itself. But this only results in judgments that are relative to the social framework within which knowledge claims are produced. If we want to raise the question of whether such claims have any validity (or any real meaning) outside this framework, we have to extend our query to the framework itself. This, however, cannot be done effectively without introducing a historical perspective, for such frameworks are historically constituted.[48] Only when we have gained some insight into the kind of historically situated reality to which a received framework has been tied can we raise the question of whether it was ever able – or will ever be able – to transcend that reality.

Addressing that type of question would take us far beyond the limited historical scope of this book. It may however be appropriate to conclude with a few hints in that direction. For example, historical analysis suggests that in the case of psychology any quantitative criterion of reality adequacy is likely to take us in the wrong direction. Thus we cannot attach any epistemic significance to the number of successfully solved empirical problems in a given domain. The one thing that modern research practices in psychology are able to do reliably is to multiply ad infinitum the number of "variables" to be investigated and hence the number of empirical studies and "solved" empirical problems. This has everything to do with the sociology and politics of research and nothing to do with progress in explicating psychological reality.[49]

A more promising procedure for addressing the question of epistemic access is likely to involve the mutual confrontation of divergent empirical domains and the different investigative practices that constitute them. This would have to include domains and practices that, for entirely extraneous reasons, have become identified with other disciplines, like linguistics, sociology, or anthropology.

However, as long as such confrontations are conducted purely as intellectual exercises they are likely to remain without practical consequences. One thing that has emerged from the historical analysis of changes in investigative practices is the importance of the social alliances formed by the discipline as a whole and by subgroups within the discipline. These alliances tended to favor and to maintain certain practices over others. An escape from methodological solipsism is likely to depend on the variety of alliances that members of the discipline manage to forge.

Working relationships and alliances are formed not only with other professional groups. We have seen that the social contexts in which psychological knowledge products are ultimately applied have an effect on the kind of knowledge product for which there is a demand and hence on the practices that must be used to produce it. In that connection it is difficult to ignore the dominant role administratively useful knowledge has played in the past. As long as that state of affairs persisted, the kind of reality to which much of psychological investigation provided access was an administratively created reality. Even when it was not directly tied to actual administrative reality, this kind of research created its own replica of such a reality, as in early American personality research.

The administrative context of application cast its shadow over significant parts of the context of investigation, which did not help to broaden the latter's access to the real world outside such contexts. The prospects of that happening would seem to depend on the extension of disciplinary alliances to groups of people who are more interested in psychological knowledge as a possible factor in their own emancipation than as a factor in their management and control of others.

The worldly success of modern psychology was built on a narrow social basis. That entailed a very considerable narrowing of epistemic access to the variety of psychological realities. Critical analysis can give us some insight into the nature of that narrowing. Further insight depends on some knowledge of that which has been excluded – in other words, knowledge that has emerged in different social contexts. The receptivity of the discipline to such knowledge, however, would seem to be tied to changes in its social and cultural commitments.

Appendix

All but one of the tables that appear in chapters 4, 5, 6, and 8 summarize parts of a content analysis of empirical articles published in psychological journals during the discipline's early years. The selection of journal volumes and articles for inclusion in this analysis was based on the following considerations.

Four basic reference points – 1895, 1910, 1925, and 1935 – were chosen for the purpose of establishing time trends. Journal volumes published during those years were included in the analysis as were the immediately preceding and the immediately following volume. For each set of three consecutive volumes of a journal, the results of the analysis were pooled to allow for minor fluctuations from one year to the next. This yielded data for the periods 1894–1896, 1909–1911, 1924–1926, and 1934–1936. There were minor discrepancies, as when the results of the analysis of the first three volumes of a journal that began publication in 1910 were included in the 1909–1911 period.

Only one of the journals analyzed, the *American Journal of Psychology,* provided a source of empirical articles over the entire period. The other journals began publication at a later stage or ceased publishing empirical studies in significant numbers at some point. The second decade of the twentieth century was a particularly fertile one in terms of the appearance of new American journals. For tracing changes in these journals the 1910 reference point was therefore impractical. Interruption of publication around 1918 introduced a further complication in some cases. To deal with these problems two procedures were adopted. For journals starting publication in this decade the first three volumes were the ones analyzed, irrespective of year of publication. For older journals the analysis of the 1909–1911 period was supplemented by an analysis of volumes published between 1914 and 1916 to reduce the time gap between volumes of these

199

journals and volumes of journals that began publication later in the decade. These data are referred to as the 1914–1916 period in tables 5.3 and 5.4, and the 1914–1920 period in table 8.1.

The following categories of journal article were not included in the analysis: theoretical articles; reviews of previously published work; articles devoted entirely to the description of apparatus; brief notes, generally not more than two pages in length; and articles based on work using only nonhuman subjects. Agreement among raters about articles to be excluded was 97 percent.

In most cases all eligible articles in a journal volume were coded. However, a random sampling of eligible articles was used in the case of the 1949–1951 time period appearing in table 5.3 and also in the case of the two applied journals for the 1924–1926 and 1934–1936 periods where coding categories were those discussed in chapter 6.

The percentages in table 6.1 are expressed in terms of the number of empirical studies using human subjects. This includes studies containing inadequate or no background information on subjects, which account for about 22 percent of all studies during the earlier periods and about 13 percent during the later periods. This would tend to depress row totals below 100. However, some studies employed more than one category of subject, thus raising row totals. The two effects tend to cancel each other out. Subjects identified as graduate students are not included in this tabulation. The category "noneducational" refers to adult subjects drawn from anywhere but an educational institution. The summary data presented here are based on an analysis of three volumes of each of six journals for the later periods, fewer for the earlier periods when some journals had not yet appeared. The journals are those identified in tables 5.2, 5.3, and 5.4.

Notes

Chapter 1. Introduction

1 To a considerable extent this is also true of physical science; see P. Janich, "Physics – natural science or technology?" in W. Krohn, E. T. Layton, Jr., and P. Weingart, eds., *The Dynamics of Science and Technology* (Dordrecht: Reidel, 1978), pp. 3–27.

2 For a general critique of the distinction between the two contexts, see P. Feyerabend, *Against Method* (London: NLB, 1975), pp. 165–169; and T. Nickles, "Scientific discovery and the future of philosophy of science," in T. Nickles, ed., *Scientific Discovery, Logic and Rationality,* Boston Studies in the Philosophy of Science, vol. 56 (Dordrecht and Boston: Reidel, 1980), pp. 1–59.

3 See S. Bem, "Context of discovery and conceptual history of psychology," in S. Bem, H. Rappard, and W. van Hoorn, eds., *Proceedings of the First European Meeting of Cheiron* (Leiden: Ryksuniversiteit, 1983), pp. 207–231.

4 "Discoveries do not simply 'occur' or 'happen' naturalistically, but are socially defined and recognized productions. In other words, the question is not what makes them happen, but rather what makes them discoveries." See A. Brannigan, *The Social Basis of Scientific Discoveries* (Cambridge: Cambridge University Press, 1981), p. 77.

5 cf. K. J. Gergen, *Toward Transformation in Social Knowledge* (New York: Springer, 1982), p. 101: "The audience for research reports is never exposed to ongoing events; one never gains first-hand experience with the research process itself. Rather, the chief product of research is language. Research reports essentially furnish linguistic accounts or interpretations."

6 A useful analysis of such constructive schemes, with illustrations from the history of experimental psychology, has been presented by Gernot Böhme, "The social function of cognitive structures: A concept of the scientific community within a theory of action," in K. D. Knorr, H. Strasser, and H. G. Zillian, eds., *Determinants and Controls of Scientific Development* (Dordrecht: Reidel, 1975), pp. 205–225; and G. Böhme, "Cognitive norms, knowledge interests and the constitution of the scientific object: A case study in the functioning of rules for experimentation," in E. Mendelsohn, P. Weingart, and R. Whitley, eds., *The Social Production of Scientific Knowledge* (Dordrecht: Reidel, 1977), pp. 129–141.

7 Psychological inquiry now acquired the general features of scientific inquiry insofar as the latter can be characterized as "a special sort of craft work which operates on intellectually constructed objects." See T. R. Ravetz, *Scientific Knowledge and Its Social Problems* (New York: Oxford University Press, 1971), p. 116.

8 It can be argued that the same situation exists in respect to any system of purely logical procedures: "As a body of conventions and esoteric traditions the compelling character of logic, such as it is, derives from certain narrowly defined purposes and from custom and institutionalized usage." See B. Barnes and D. Bloor, "Relativism, rationalism and the sociology of knowledge," in M. Hollis and S. Lukes, eds., *Rationality and Relativism* (Oxford: Blackwell, 1982), pp. 21–47 (p. 45).

9 G. Devereux, *From Anxiety to Method in the Behavioral Sciences* (The Hague: Mouton, 1967). Devereux was the first to recognize the depth of anxiety surrounding the area of methodology as an indication of fundamental problems in the behavioral sciences. However, he saw the sources of this anxiety only in terms of problems arising for the individual investigator in his relationship to the human subject of his investigation. The problems of investigative practice therefore appeared analogous to the problems of psychoanalytic practice. This radically individualist framework produced a complete blindness toward the collective anxieties of groups of investigators arising out of real or imagined threats to their position in the scientific status structure.

10 Some important consequences are discussed in R. D. Romanyshyn, "Method and meaning in psychology: The method has been the message," *Journal of Phenomenological Psychology* 2 (1971):93–113; and C. Argyris, *Inner Contradictions of Rigorous Research* (New York: Academic Press, 1980).

11 The scope of the profound changes that have occurred in this field is perhaps most readily appreciated by the outsider through a comparison of a text representing the old approach, like J. Ben-David, *The Scientist's Role in Society* (Englewood Cliffs, N.J.: Prentice-Hall, 1971), and one informed by more recent developments, like J. Law and P. Lodge, *Science for Social Scientists* (London: Macmillan, 1984). For a succinct introduction to issues in this field, see the introduction to K. D. Knorr-Cetina and M. Mulkay, eds., *Science Observed: Perspectives on the Social Study of Science* (London and Beverly Hills, Calif.: Sage, 1983), pp. 1–17.

12 E.g., A. R. Buss, ed., *Psychology in Social Context* (New York: Irvington, 1979).

13 The existence of a long-unheeded exception drives home its complete isolation in the literature; see S. Rosenzweig, "The experimental situation as a psychological problem," *Psychological Review* 40 (1933):337–354. Because of their exposure to real world problems, industrial psychologists were forced to recognize social features in investigative situations at an early stage; see W. Schulte, "Untersuchungen über den Einfluss des Versuchsleiters auf das Prüfergebnis," *Industrielle Psychotechnik* 1 (1924):289–291; and Elton Mayo, *The Human Problems of an Industrial Civilization* (New York: Macmillan, 1933). However, the extension of these insights from the "profane" area of industrial settings to the "sacred" precincts of university laboratories seems to have encountered formidable cognitive and affective barriers. A parallel case occurred in the field of animal experimentation; see D. Fernald, *The Hans Legacy: A Story of Science* (Hillsdale, N.J., and London: Erlbaum, 1984).

14 Discussions of this literature are found in T. X. Barber, *Pitfalls in Human Research: Ten Pivotal Points* (New York: Pergamon Press, 1976); R. Rosenthal, *Experimenter Effects in Behavioral Research*, rev. ed. (New York: Irvington, 1976); R. Rosenthal and R. L. Rosnow, eds., *Artifact in Behavioral Research* (New York:

Academic Press, 1969); R. Rosenthal and R. L. Rosnow, *The Volunteer Subject* (New York: Wiley, 1975); R. Rosenthal and D. B. Rubin, "Interpersonal expectancy effects: The first 345 studies," *The Behavioral and Brain Sciences* 3 (1978):377–415; J. M. Suls and R. L. Rosnow, "Concerns about artifacts in psychological experiments," in J. G. Morawski, ed., *The Rise of Experimentation in American Psychology* (New Haven: Yale University Press, 1988), pp. 163–187.

15 J. G. Adair, "Commentary," *The Behavioral and Brain Sciences* 3 (1978):386.

16 R. M. Farr, "On the social significance of artifacts in experimenting," *British Journal of Social and Clinical Psychology* 17 (1978):299–306.

17 L. W. Brandt, "Experimenter-effect research," *Psychologische Beiträge* 17 (1975):133–140.

18 See, however, J. G. Adair, *The Human Subject: The Social Psychology of the Psychological Experiment* (Boston: Little, Brown, 1973); and N. Friedman, *The Social Nature of Psychological Research: The Psychological Experiment as a Social Interaction* (New York: Basic Books, 1967).

19 D. P. Schultz, "The nature of the human data source in psychology," in D. P. Schultz, ed., *The Science of Psychology: Critical Reflections* (New York: Appleton-Century-Crofts, 1970), pp. 77–86.

20 K. Holzkamp, *Kritische Psychologie* (Frankfurt: Fischer, 1972); W. Mertens, *Sozialpsychologie des Experiments* (Hamburg: Hoffman & Campe, 1975); W. Bungard, ed., *Die gute Versuchsperson denkt nicht: Artefakte in der Sozialpsychologie* (Munich and Baltimore: Urban & Schwarzenberg, 1980).

21 H. M. Collins, *Changing Order: Replication and Induction in Scientific Practice* (London and Beverly Hills, Calif.: Sage, 1985). See also K. Knorr, R. Krohn, and R. Whitley, eds., *The Social Process of Scientific Investigation, Sociology of the Sciences Yearbook,* vol. 4 (Boston and Dordrecht: Reidel, 1980); and K. D. Knorr-Cetina, *The Manufacture of Knowledge: An Essay on the Constructivist and Contextual Nature of Science* (Oxford: Pergamon Press, 1981).

22 Moreover, there are inherent limitations on the use of experimental methods to investigate the experimental method; see H. Gadlin and G. Ingle, "Through the one-way mirror: The limits of experimental self-reflection," *American Psychologist* 30 (1975):1003–1009.

23 R. Whitley, *The Intellectual and Social Organization of the Sciences* (Oxford: Clarendon Press, 1984).

24 T. M. Ziman, *Public Knowledge: An Essay concerning the Social Dimension of Science* (Cambridge: Cambridge University Press, 1968).

25 E. B. Titchener, "Psychology: Science or technology?" *Popular Science Monthly* 84 (1914):39–51.

26 Because of their pedagogical mission, textbook histories frequently suggest some such position; see M. G. Ash, "The self-presentation of a discipline: History of psychology in the United States between pedagogy and scholarship," in L. Graham, W. Lepenies, and P. Weingart, eds., *Functions and Uses of Disciplinary Histories, Sociology of the Sciences Yearbook,* vol. 7 (Dordrecht: Reidel, 1983), pp. 143–189.

27 On the justificationist implications of this position, see W. B. Weimer, "The history of psychology and its retrieval from historiography: I. The problematic nature of history," *Science Studies* 4 (1974):235–258.

28 This recognition depended on an emphasis on the craft character of science by M. Polyani, *Personal Knowledge* (London: Routledge and Kegan Paul, 1958). The implications of such an analysis were extensively developed by Ravetz, *Scientific Knowledge*.

29 These kinds of studies have proliferated greatly in recent years. A well-known early example of the observational study of scientific work was B. Latour and S. Woolgar, *Laboratory Life: The Social Construction of Scientific Facts* (London and Beverly Hills, Calif.: Sage, 1979). The use of interview material is exemplified by G. N. Gilbert and M. Mulkay, *Opening Pandora's Box: A Sociological Analysis of Scientists' Discourse* (Cambridge: Cambridge University Press, 1984).

30 For some sociologists of science the existence of this distinction appears to have provided the occasion for claiming a privileged status for either the direct observation of practice or the analysis of discourse about practice. The latter position has been forcefully argued in G. N. Gilbert and M. Mulkay, "Experiments are the key: Participants' histories and historians' histories of science," *Isis* 75 (1984):105–125 (but see S. Shapin, "Talking history: Reflections on discourse analysis," *Isis* 75 [1984]:125–130). However, the production of a certain kind of discourse in the form of scientific publications lies at the core of scientific practice itself. If we wish to follow the historical development of science, we should be looking at this discourse. But because this discourse is part of the ordinary practice of science, the useful distinction to make here is that between unreflected practice and reflections on that practice, not between "discourse" on the one hand and actions and beliefs on the other.

31 A good example of the use of scientific articles as historical documents is provided by C. Bazerman, "Modern evolution of the experimental report in physics: Spectroscopic articles in *Physical Review,* 1893–1980," *Social Studies of Science* 14 (1984):163–196. See also C. Bazerman, *Shaping Written Knowledge: The Genre and Activity of the Written Article in Science* (Madison: University of Wisconsin Press, 1988).

32 Two major exceptions are J. S. Bruner and G. W. Allport, "Fifty years of change in American psychology," *Psychological Bulletin* 37 (1940):757–776; and R. W. Lissitz, "A longitudinal study of the research methodology in the Journal of Abnormal and Social Psychology, the Journal of Nervous and Mental Disease and the American Journal of Psychiatry," *Journal of the History of the Behavioral Sciences* 5 (1969):248–255.

33 A comprehensive review of the use of the second type of analysis is provided by C. Bazerman, "Scientific writing as a social act: Review of the literature of the sociology of science," in P. V. Anderson, R. J. Brockmann, and C. R. Miller, eds., *New Essays in Technical and Scientific Communication: Research, Theory, Practice* (Farmingdale, N.Y.: Baywood, 1983), pp. 156–184. Once the informal rules governing the presentation of the knowledge product have become formalized they can also be directly analyzed. See C. Bazerman, "Codifying the social scientific style: The APA Publication Manual as a Behaviorist Rhetoric," in J. S. Nelson, A. Megill, D. N. McCloskey, eds., *The Rhetoric of the Human Sciences* (Madison: University of Wisconsin Press, 1987), pp. 125–144.

34 On the relationship between research publications and research practices, see F. L. Holmes, "Scientific writing and scientific discovery," *Isis* 78 (1987):220–235.

35 See S. Toulmin and D. E. Leary, "The cult of empiricism in psychology and beyond," in S. Koch and D. E. Leary, eds., *A Century of Psychology as Science: Retrospectives and Assessments* (New York: McGraw-Hill, 1985), pp. 594–617.

36 This issue is explored further in K. Danziger, "Towards a conceptual framework for a critical history of psychology," in H. Carpintero and J. M. Peiro, eds.,

Psychology in Its Historical Context: Essays in Honour of Prof. Josef Brozek (Valencia: Monografias de la Revista de Historica de la Psicologia, 1984), pp. 99–107.

Chapter 2. Historical roots of the psychological laboratory

1 See W. G. Bringmann, N. J. Bringmann, and G. A. Ungerer, "The establishment of Wundt's laboratory: An archival and documentary study," in W. G. Bringmann and R. D. Tweney, eds., *Wundt Studies: A Centennial Collection* (Toronto: C. J. Hogrefe, 1980), pp. 123–157.

2 Empiricist mental philosophy did of course proceed in a methodical way from complex ideas to "simple" ones that had to be isolated in experience, and this according to Locke "requires pains and assiduity" (*An Essay concerning Human Understanding,* bk. 2, chap. 13, sec. 28), but this was a general prescription, not related to the special problems of the "inner sense."

3 The closest that even an archintrospectionist like James Mill comes to methodological self-reflection is when he advises that "the learner should by practice acquire the habit of reflecting upon his Sensations, as a distinct class of feelings." See J. Mill, *Analysis of the Phenomena of the Human Mind* (1829; repr. New York: A. M. Kelley, 1967), p. 2.

4 The most important examples are provided by John Stuart Mill's defense of introspection in his *Auguste Comte and Positivism* (London: G. Routledge, [1865]), and William Hamilton's *Lectures on Metaphysics and Logic* (Edinburgh: Blackwood, 1859), lecture 19, which is strongly influenced by his study of Kant.

5 Immanuel Kant, *Die metaphysischen Anfangsgründe der Naturwissenschaft* (1786); see esp. p. 8 of the English translation, *Metaphysical Foundations of Natural Science,* trans. J. Ellington (Indianapolis: Bobbs-Merrill, 1970). To get the measure of Kant's new methodological sensitivity one need only compare his treatment of the evidential value of the inner sense with that of his guide in matters psychological, Johann Nicolas Tetens. See J. N. Tetens, *Philosophische Versuche über die menschliche Natur und ihre Entwicklung* (Leipzig: Weidmann u. Reich, 1777). Tetens in turn shows more awareness of methodological problems than his predecessors.

6 For a good modern discussion of this issue, see R. Rorty, *Philosophy and the Mirror of Nature* (Princeton: Princeton University Press, 1979), chap. 3, sec. 2.

7 On Kant and psychology, see T. Mischel, "Kant and the possibility of a science of psychology," *Monist* 51 (1967):599–622; C. Gouaux, "Kant's view on the nature of empirical psychology," *Journal of the History of the Behavioral Sciences* 8 (1972):237–242; D. E. Leary, "Immanuel Kant and the development of modern psychology," in W. R. Woodward and M. G. Ash, eds., *The Problematic Science: Psychology in Nineteenth Century Thought* (New York: Praeger, 1982), pp. 17–42; and G. Verwey, *Psychiatry in an Anthropological and Biomedical Context* (Dordrecht: D. Reidel, 1985).

8 Kant's *Anthropologie in pragmatischer Hinsicht* (1798) gives a good idea of what such a psychology would look like. See the translation *Anthropology from a Pragmatic Point of View,* by M. J. Gregor (The Hague: Martinus Nijhoff, 1974).

9 See B. D. Mackenzie and S. L. Mackenzie, "The case for a revised systematic approach to the history of psychology," *Journal of the History of the Behavioral Sciences* 10 (1974):324–347. The argument is based on E. A. Burtt, *The Metaphysical Foundations of Modern Physical Science* (London: Routledge & Kegan Paul, 1932), and A. Koyré, *From the Closed World to the Infinite Universe* (Baltimore: Johns Hopkins University Press, 1957).

10 Edmund Husserl gives a succinct description of this process in his *The Crisis of European Sciences and Transcendental Phenomenology,* pt. 2 (Evanston, Ill.: Northwestern University Press, 1970).

11 See Rorty, *Mirror of Nature.*

12 J. F. Herbart, *Psychologie als Wissenschaft, neu gegründet auf Erfahrung, Metaphysik und Mathematik* (Königsberg: Unger, 1824); also, *Ueber die Möglichkeit und Nothwendigkeit, Mathematik auf Psychologie anzuwenden,* vol. 5 of *Sämtliche Werke,* ed. K. Kehrbach (Langensalza: Beyer, 1890).

13 Kant, *Anthropology,* p. 22.

14 The major German representatives of this latter trend were, in chronological order, J. F. Fries, F. E. Beneke, and A. Fortlage. See also D. E. Leary, "The philosophical development of the conception of psychology in Germany 1780–1880," *Journal of the History of the Behavioral Sciences* 14 (1978):113–121. None of these figures received much recognition from the dominant German philosophical circles, and Beneke suffered political persecution for his views.

15 In France the philosophical accents were somewhat different, but in Auguste Comte one meets the same combination of distrust of introspection and affirmation of supraindividual principles of order.

16 This was of course the prevailing orthodoxy in nineteenth-century British empiricist philosophy, in the Scottish school of philosophy, and in early American psychology. One reason for the revolutionary impact of behaviorism in the American context must certainly be traced to its provocative break with the received basis of Anglo-Saxon mental philosophy.

17 After about 1860 the attack on introspection as a method is typically grounded in a mixture of traditional philosophical arguments and more modern concerns for the claims of a supraindividual order. Leading examples of the new trend are the German neo-Kantian socialist F. A. Lange (see his *The History of Materialism,* vol. 2[1865], chap. 3), and the spokesman of British institutional psychiatry, Henry Maudsley (see his *The Physiology of Mind* [1867], chap. 1). Lange coined the slogan "psychology without the soul." Both he and Maudsley played an important role in popularizing the idea of a scientific psychology in the years preceding the establishment of Wundt's laboratory at Leipzig. On Lange's interesting biography, see O. A. Ellissen, *Friedrich Albert Lange: Eine Lebensbeschreibung* (Leipzig: Baedeker, 1891).

18 See D. Bloor, *Wittgenstein: A Social Theory of Knowledge* (New York: Columbia University Press, 1983), chap. 4.

19 Hacking, for instance, distinguishes between the test, the adventure, the diagnosis, and the dissection. See I. Hacking, *The Emergence of Probability* (Cambridge: Cambridge University Press, 1975), pp. 35–37.

20 W. Wundt, *Grundzüge der physiologischen Psychologie* (Leipzig: Engelmann, 1874). In the introduction to this crucial work Wundt defines the new approach as follows: "Psychological introspection goes hand in hand with the methods of experimental physiology. If one wants to put the main emphasis on the characteristic of the method, our science, experimental psychology, is to be distinguished from the ordinary mental philosophy [*Seelenlehre*] based purely on introspection" (pp. 2–3).

21 Because English physiology was still a few years away from this development, there could be no question of it nurturing the emergence of an experimental psychology at that time. See G. L. Geison, "Social and institutional factors in the stagnancy of English physiology, 1840–1870," *Bulletin of the History of Medicine* 46 (1972):30–58.

22 I speak of the hopes of the *young* Wundt advisedly because in due course he lost the enthusiasm for the possibilities of the experimental method in psychology that he had expressed in his first major publication. See the introduction to his *Beiträge zur Theorie der Sinneswahrnehmung* (Leipzig and Heidelberg: Winter, 1862).

23 J. M. D. Olmsted, *François Magendie* (New York: Schuman's, 1944); O. Temkin, "Basic science, medicine and the romantic era," *Bulletin of the History of Medicine* 37 (1963):97–129 (p. 121); see also P. F. Cranefield, *The Way in and the Way out: François Magendie, Charles Bell and the Roots of the Spinal Nerves* (Mt. Kisco, N.Y.: Futura, 1974).

24 There are some interesting parallels between the history of the institutional relationship of anatomy and physiology and of philosophy and psychology. Both relationships were transformed when the junior partner redefined its subject matter in terms of the kind of knowledge yielded by the practice of experimentation.

25 See J. Schiller, "Physiology's struggle for independence in the first half of the nineteenth century," *History of Science* 7 (1968):64–89.

26 For a detailed analysis of these changes, and the connection between functionalism and experimentalism, see W. R. Albury, "Experiment and explanation in the physiology of Bichat and Magendie," *Studies in History of Biology* 1 (1977):47–132; a different view of the relation between Bichat and Magendie is presented in J. E. Lesch, *Science and Medicine in France* (Cambridge, Mass.: Harvard University Press, 1984). On the significance of the functional perspective, see also M. Foucault, *The Order of Things* (London: Tavistock, 1970), pp. 228 and 264.

27 See W. Coleman, "The cognitive basis of the discipline: Claude Bernard on physiology," *Isis* 76 (1985):49–70.

28 I have analyzed this development more fully elsewhere. See K. Danziger, "Origins of the schema of stimulated motion: Towards a prehistory of modern psychology," *History of Science* 21 (1983):183–210.

29 This received its most influential textbook codification in Johannes Müller's, *Handbuch der Physiologie des Menschen* (Coblenz: Holscher, 1833–1840). An English translation appeared in 1842. For slightly later English language developments in this direction, see K. Danziger, "Mid-nineteenth century British psychophysiology: A neglected chapter in the history of psychology," in W. R. Woodward and M. G. Ash, eds., *The Problematic Science: Psychology in Nineteenth Century Thought* (New York: Praeger, 1982), pp. 119–146.

30 The practical effect was mediated by the Russian physiological school of Sechenov and Pavlov and never became part of the mainstream of western psychology.

31 The classical example is of course Fechner's appropriation of E. H. Weber's experiments in sensory physiology with the purpose of providing some empirical foundation for his metaphysical intuitions about the soul. See M. E. Marshall, "Physics, metaphysics, and Fechner's psychophysics," in Woodward and Ash, *The Problematic Science,* pp. 65–87. This line of thought had been made possible, among other things, by the fact that the new experimental physiology forced a relocalization of sensory experience in the central parts of the nervous system, the same parts that were often regarded as the seat of the soul. See M. Gross, "The lessened locus of feelings: A transformation in French physiology in the early nineteenth century," *Journal of the History of Biology* 12 (1979):231–271.

32 Wundt, *Grundzüge.*

33 From Wundt's own statistical summary of the experimental publications emanating from his laboratory during its first quarter century, it appears that three-quarters of all these studies were of this type. Moreover, most of the others fell into the latter part of this time period. W. Wundt, "Das Institut für experimentelle Psychologie," in *Die Institute und Seminare der philosophischen Fakultät an der Universität Leipzig* (Leipzig: Hirzel, 1909), pp. 118–133.

34 For an incisive analysis of the principles involved in the practices of this first community of scientific experimenters, see S. Shapin and S. Schaffer, *Leviathan and the Air-Pump: Hobbes, Boyle, and the Experimental Life* (Princeton: Princeton University Press, 1985).

35 See Bringmann et al., "Wundt's laboratory"; and Wundt, "Das Institut."

36 For comprehensive statistics on this expansion, see W. Lexis, *Die deutschen Universitäten* (Berlin: Asher, 1893); also J. Conrad, "Ergebnisse der deutschen Universitätsstatistik," *Jahrbücher für Nationalökonomie und Statistik* 32 (1906): 433–492. Wundt himself presents an interesting analysis of the local factors that favored the success of his venture at the University of Leipzig; see W. Wundt, *Festrede zur funfhundertjährigen Jubelfeier der Universität Leipzig* (Leipzig: Engelmann, 1909).

37 M. A. Tinker, "Wundt's doctorate students and their theses," in Bringmann and Tweney, *Wundt Studies,* pp. 269–279; also S. Fernberger, "Wundt's doctorate students," *Psychological Bulletin* 30 (1933):80–83.

38 In the United States Titchener's laboratory at Cornell was particularly important in this respect. See R. D. Tweney, "Programmatic research in experimental psychology: E. B. Titchener's laboratory investigations, 1891–1927," in M. G. Ash and W. R. Woodward, eds., *Psychology in Twentieth-Century Thought and Society* (Cambridge: Cambridge University Press, 1987), pp. 35–57.

39 I have adopted the use of the term "technology" in this context and the distinction between material, literary and social technologies from Shapin and Schaffer, *Leviathan.* These are analytical distinctions; in the real world the different technologies are closely interwoven.

40 For the history of these forms, see C. E. McClelland, *State, Society, and University in Germany, 1700–1914* (Cambridge: Cambridge University Press, 1980), pp. 174–181.

41 The University of Berlin illustrates the trend: In 1811, 72 percent of its budget went on salaries and 24 percent on institutes; by 1896 salaries accounted for only 31 percent and institutes for 53 percent; see F. Paulsen, *Die deutschen Universitäten und das Universitätsstudium* (Berlin: Asher, 1902).

42 With the growth of the institute the organization of the work of its members began to assume distinctly industrial features. In 1893 Wundt describes an organization with coordinated work teams, team leaders, and planned division of topics. By then the institute apparently had twenty-five members; see Wundt, "Psychophysik und experimentelle Psychologie," in Lexis, *Die deutschen Universitäten,* pp. 454–455.

43 The most important of these were the chronoscope and the kymograph. For recently republished illustrations of these and other pieces of apparatus used in Wundt's laboratory, see J. A. Popplestone and M. White McPherson, "The vitality of the Leipzig model of 1880–1910 in the United States in 1950–1980," in Bringmann and Tweney, *Wundt Studies,* pp. 226–257; also W. Traxel, H. Gundlach, and U. Zschuppe, "Zur Geschichte der apparativen Hilfsmittel der Psychologie," in R. Brickenkamp, ed., *Handbuch der apparativen Verfahren der Psychologie* (Göttingen: Hogrefe, 1986), pp. 1–22. See also M. Borell, "Instrumentation and the rise of physiology," *Science and Technology Studies* 5 (1987):53–62.

44 At the beginning of his experimental program Wundt declared that experimental methods would fail in psychology "whenever an intelligent consent [*verständnisvolles Eingehen*] to the intentions of the psychologist could not be assumed" (p. 208). W. Wundt, "Die Aufgaben der experimentellen Psychologie" (1882) in *Essays*, 2d ed. (Leipzig: Engelmann, 1906), pp. 187–212.

45 See M. Sokal, ed., *An Education in Psychology: James McKeen Cattell's Journal and Letters from Germany and England 1880–1888* (Cambridge, Mass.: MIT Press, 1980).

46 See the papers by Cattell, Friedrich, Merkel, Starke, Tischer, and Trautscholdt in the first three volumes of the journal *Philosophische Studien*, 1883–1886.

47 This may be an example of the process of *typification* described by P. Berger and T. Luckmann, *The Social Construction of Reality* (Harmondsworth: Penguin, 1967).

48 J. Kollert, "Untersuchungen über den Zeitsinn," *Philosophische Studien* 1 (1883):78–89; V. Estel, "Neue Versuche über den Zeitsinn," *Philosophische Studien* 2 (1885):37–65.

49 E.g., J. Merkel, "Die zeitlichen Verhältnisse der Willensthätigkeit," *Philosophische Studien* 2 (1885):73–127. The same lack of consistency can be observed in the earliest English-language experimental reports; cf. the papers by G. S. Hall in *Mind*, vols. 10 and 11, and *American Journal of Psychology*, vol. 1.

50 The equipment of Wundt's laboratory was not only copied; an industrial offshoot of the laboratory supplied such equipment to adequately funded customers. Wundt's mechanic, Ernst Zimmermann, established his own firm in 1887 and produced psychological apparatus in series. This development certainly assisted the standardization of psychological research on an international scale. Subsequently, the firm of C. H. Stoelting in Chicago fulfilled a rather similar function. See Traxel, Gundlach and Zschuppe, "Hilfsmittel der Psychologie."

Chapter 3. Divergence of investigative practice:
The repudiation of Wundt

1 W. Wundt, "Die Aufgaben der experimentellen Psychologie" (1882), in *Essays*, 2d ed. (Leipzig: Engelmann, 1906), pp. 187–212.

2 See Franz Brentano, *Psychologie vom empirischen Standpunkt*, vol. 1 (1874; repr. Hamburg: Meiner, 1955), chap. 2. Although he does not acknowledge any indebtedness, Wundt was clearly aware of Brentano's distinction and may well have decided to adopt it. Of course, the essential point goes back to Kant.

3 It is most unfortunate that English-language references to Wundt's position have so often failed to reproduce this basic distinction and have generally used the term "introspection" to cover both concepts indiscriminately. This of course makes it impossible to understand Wundt's practice.

4 W. Wundt, "Selbstbeobachtung und innere Wahrnehmung," *Philosophische Studien* 4 (1887):292–309. For a concise account of Wundt's conception of the relation of introspection and experiment at the time of the founding of his laboratory, see also his *Methodenlehre* of 1883 (pp. 282ff.). (This was the second volume of *Logik* [Stuttgart: Enke, 1883].)

5 A major function of experimental arrangements was the production of *precise* reports. These arrangements "force introspection to give an answer to a precisely put question;" see W. Wundt, "Hypnotismus und Suggestion," *Philosophische Studien* 8 (1893):1–85 (p. 65).

6 See, e.g., W. Wundt, "Ueber psychologische Methoden," *Philosophische Studien* 1 (1883):1–40.

7 Early versions of such exclusions are to be found in the first edition of Wundt's *Vorlesungen über die Menschen- und Thierseele* (Leipzig: Voss, 1863). Subsequently, the theme is developed in successive editions of his *Logik,* in his writings on *Völkerpsychologie,* and in his "Ueber Ausfrageexperimente und über die Methoden zur Psychologie des Denkens," *Psychologische Studien* 3 (1907):301–360.

8 For an overview, see K. Danziger, "Origins and basic principles of Wundt's *Völkerpsychologie,*" *British Journal of Social Psychology* 22 (1983):303–313.

9 See the preface to the third edition of Wundt's *Logik.*

10 Cf. W. van Hoorn and T. Verhave, "Wundt's changing conceptions of a general and theoretical psychology," in W. G. Bringmann and R. D. Tweney, eds., *Wundt Studies: A Centennial Collection* (Toronto: C. J. Hogrefe, 1980), pp. 71–113.

11 For a more detailed discussion of this topic, see K. Danziger, "Wundt's psychological experiment in the light of his philosophy of science," *Psychological Research* 42 (1980):109–122.

12 W. Wundt, *Grundzüge der physiologischen Psychologie,* 5th ed., vol. 3 (Leipzig: Engelmann, 1903), p. 703.

13 For a version that is available in English, see part 5 of Wundt's *Outlines of Psychology* (1897), reprinted in R. W. Rieber, ed., *Wilhelm Wundt and the Making of a Scientific Psychology* (New York: Plenum, 1980), pp. 179–195; see also T. Mischel, "Wundt and the conceptual foundations of psychology," *Philosophical and Phenomenological Research* 31 (1970):1–26.

14 Wundt neither was nor referred to himself as a "structuralist." This term was invented by Titchener for the purpose of distinguishing his own approach from that of many of the American psychologists. Some of the fundamental differences between Titchener and Wundt are discussed in the next section. For an overview of Wundt's position, see A. Blumenthal, "Wilhelm Wundt: Psychology as the propadeutic science," in C. E. Buxton, ed., *Points of View in the Modern History of Psychology* (New York: Academic Press, 1985), pp. 19–49; K. Danziger, "Wundt and the two traditions of psychology," and "Wundt's theory of behavior and volition," in Rieber, *Wundt and Scientific Psychology,* pp. 73–115; W. R. Woodward, "Wundt's program for the new psychology: Vicissitudes of experiment, theory and system," in W. R. Woodward and M. G. Ash, eds., *The Problematic Science: Psychology in Nineteenth Century Thought* (New York: Praeger, 1980), pp. 167–197.

15 See F. K. Ringer, *The Decline of the German Mandarins* (Cambridge, Mass.: Harvard University Press, 1969).

16 Among the more important of these contributions were Wundt's *Ethik* (1886) and his *System der Philosophie* (1889), both published at Leipzig by Engelmann.

17 Cf. A. Métraux, "Wilhelm Wundt und die Institutionalisierung der Psychologie," *Psychologische Rundschau* 31 (1980):84–98.

18 Wundt's work must be understood in terms of the complex and fluid relationships between philosophy and science (and philosophy and psychology) in nineteenth-century German academic life. For a good modern overview of some of this background, see H. Schnädelbach, *German Philosophy 1831–1933* (Cambridge: Cambridge University Press, 1983).

19 On G. E. Müller, see A. L. Blumenthal, "Shaping a tradition: Experimentalism begins," in Buxton, *Modern History of Psychology,* chap. 3, pp. 51–83.

20 For further discussion of this concept, see chapter 8.

21 W. Wundt, "Die Psychologie im Kampf ums Dasein" (1913), in his *Kleine Schriften,* vol. 3 (Stuttgart: Kröner, 1921), pp. 515–543.

22 In his 1877 review of the state of philosophy in Germany, written for English-speaking readers, Wundt presents statistics on the number of lecture series devoted to the various subfields of philosophy at all the German universities during the period 1874–1877. The number of lecture courses in psychology exceeded those in metaphysics by a factor of about three and those in ethics by a factor of about four. However, logic was slightly more popular than psychology. See W. Wundt, "Philosophy in Germany," *Mind* 2 (1877):493–518.

23 Cf. Windelband's caustic comment on the situation: "For a time things in Germany had almost got to the point where the ability to occupy a chair in philosophy was regarded as established once someone had learned to push electrical buttons in a methodical manner." W. Windelband, *Die Philosophie im deutschen Geistesleben des 19. Jahrhunderts* (Tübingen: Mohr, 1909), p. 92.

24 For a comprehensive account of this process, see M. G. Ash, "Wilhelm Wundt and Oswald Külpe on the institutional status of psychology: An academic controversy in historical context," in Bringmann and Tweney, *Wundt Studies,* pp. 396–421; and M. G. Ash, "Academic politics in the history of science: Experimental psychology in Germany, 1879–1941," *Central European History* 13 (1981):255–286. W. Wirth, one of the post-Wundtian generation of German experimental psychologists who was fairly close to Wundt's position, refers to being victimized by the "philosophic countercurrent" in his autobiographical statement in C. Murchison, *A History of Psychology in Autobiography,* vol. 3 (Worcester, Mass.: Clark University Press, 1936), pp. 283–327 (p. 292).

25 For further discussion of these differences, see chapter 8, and K. Danziger, "The social origins of modern psychology," in A. R. Buss, ed., *Psychology in Social Context* (New York: Irvington, 1979), pp. 27–45.

26 For a more detailed analysis of this development, see K. Danziger, "The positivist repudiation of Wundt," *Journal of the History of the Behavioral Sciences* 15 (1979):205–230.

27 The major philosophical underpinning for these positions were provided by Ernst Mach and R. H. L. Avenarius whose writings exerted a strong influence on some of the key figures among the generation of experimental psychologists that immediately followed Wundt. The latter attacked this position in a series of long papers in *Philosophische Studien* between 1894 and 1898. For details, see Danziger, "The Positivist Repudiation of Wundt."

28 E. B. Titchener, "Prolegomena to a study of introspection," *American Journal of Psychology* 23 (1912):427–448 (p. 427). See also Titchener, "The problems of experimental psychology," *American Journal of Psychology* 16 (1905):208–224.

29 Titchener himself, Alfred Binet, and Theodor Lipps should probably be counted as the most prominent non-Würzburg proponents of a less restrictive use of introspection in an experimental context. On the Würzburg experiments, see G. Humphrey, *Thinking* (London: Methuen, 1950).

30 W. Wundt, "Ueber Ausfrageexperimente und über die Methoden zur Psychologie des Denkens," *Psychologische Studien* 3 (1907):301–360.

31 My account of these changes was originally published in K. Danziger, "The history of introspection reconsidered," *Journal of the History of the Behavioral Sciences* 16 (1980):241–262.

32 The most explicit advocate of these methods was Narziss Ach, who coined the term "systematic experimental introspection." See N. Ach, *Ueber die Willenstätigkeit und das Denken* (Göttingen: Vandenhoeck und Ruprecht, 1905).

33 Cf. Ach, ibid., and J. Segal, "Ueber den Reproduktionstypus und das Reproduzieren von Vorstellungen," *Archiv für die gesamte Psychologie* 12 (1908):124–235.

34 G. E. Müller, *Zur Analyse der Gedächtnistätigkeit und des Vorstellungsverlaufes* (1911; repr., Leipzig: Barth, 1924).

35 F. Galton, *Inquiries into Human Faculty and Its Development* (London: Macmillan, 1883).

36 A. Binet. "La mésure de la sensibilité," *L'Année psychologique* 9 (1903): 79–128.

37 These studies are discussed in more detail in chapter 9 of the present volume.

38 See Shapin and Schaffer, *Leviathan.*

39 The plea for the reorientation of psychological research toward the production of knowledge useful for human management was central to Watson's behaviorist manifesto: "If psychology would follow the plan I suggest the educator, the physician, the jurist and the businessman could utilize our data in a practical way, as soon as we are able, experimentally, to obtain them." J. B. Watson, "Psychology as the behaviorist views it," *Psychological Review* 20 (1913):158–177. Note the stark appeal for a change of disciplinary project: "*If psychology would follow the plan I suggest.*"

40 Although Titchener would not have wished to equate the viewpoint of psychology with that of physical science, his strictures on the language of introspection represented a kind of reductio ad absurdum of the neutral, value-free, unambiguous language of scientific description. He insisted on the elimination, not only of value and ambiguity, but of meaning itself. The classical statement of his position is to be found in E. B. Titchener, "Description vs. statement of meaning," *American Journal of Psychology* 23 (1912):165–182; also see his "The schema of introspection" in the same volume, pp. 485–508.

41 E. von Aster, "Die psychologische Beobachtung und experimentelle Untersuchung von Denkvorgängen," *Zeitschrift für Psychologie* 49 (1908):56–107.

42 Essentially the same distinction as that made by von Aster was also made independently, though less elegantly, by Dürr, a competent experimenter who had himself been a subject in Bühler's Würzburg experiments. See E. Dürr, "Ueber die experimentelle Untersuchung der Denkvorgänge," *Zeitschrift für Psychologie* 49 (1908):315. Titchener accepted his own version of the distinction in which the place of *Kundgabe* is taken by "information." Characteristically, this was an intellectualistic version of the original distinction in which description is now contrasted with interpretation and not with expression, as was the original intention. As usual, the spirit of Titchener survived in his student; see E. G. Boring's treatment of the issue in *A History of Experimental Psychology* (New York: Appleton-Century-Crofts, 1950), p. 610. After a gap of several decades there has been a more recent revival of interest in the topic of introspection. See W. Lyons, *The Disappearance of Introspection* (Cambridge, Mass.: MIT Press, 1986).

43 Von Aster, "Die psychologische Beobachtung."

Chapter 4. The social structure of psychological experimentation

1 R. M. Farr, "Social representations: Their role in the design and execution of laboratory experiments," in R. M. Farr and S. Moscovici, eds., *Social Representations* (Cambridge: Cambridge University Press, 1984).

2 M. Mehner, "Zur Lehre vom Zeitsinn," *Philosophische Studien* 2 (1885):546–602.

3 For example, Wundt appears as subject in the following papers published by his students in the first volume of the journal *Philosophische Studien* (1883–1884): M. Friedrich, "Ueber die Apperceptionsdauer bei einfachen und zusammengesetzten Vorstellungen," pp. 39–77; J. Kollert, "Untersuchungen über den Zeitsinn," pp. 78–89; M. Trautscholdt, "Experimentelle Untersuchungen über die Association der Vorstellungen," pp. 213–250; E. Tischer, "Ueber die Unterscheidung von Schallstärken," pp. 495–542.

4 The case of two of Wundt's students, Cattell and Berger, has been documented in some detail on the basis of Cattell's letters in M. Sokal, ed., *An Education in Psychology: James McKeen Cattell's Journals and Letters from Germany and England, 1880–1888* (Cambridge, Mass.: MIT Press, 1980). Another early instance of a pair of investigators who alternated as experimenter and subject is provided by Lorenz and Merkel; see G. Lorenz, "Die Methode der richtigen und falschen Fälle in ihrer Anwendung auf Schallempfindungen," *Philosophische Studien* 2 (1885): 394–474.

5 See, e.g., L. Lange, "Neue Experimente über den Vorgang der einfachen Reaction auf Sinneseindrücke," *Philosophische Studien* 4 (1888):497–510; and E. B. Titchener, "Zur Chronometrie des Erkennungsactes," *Philosophische Studien* 8 (1893):138–144.

6 For a succinct insider's account of the relevant institutional practices, see the translation of a late nineteenth-century German guidebook, F. Paulsen, *The German Universities and University Study* (New York: Scribner, 1906), p. 212: "The student can be introduced to scientific research only by the method of cooperation. And that is the real purpose of the seminars; they are the nurseries of scientific research. In them, under the guidance and assistance of a master, pupils become acquainted with scientific work and learn how to do it. After their apprenticeship they continue to work themselves as masters." On the significance of the German university system for the organization of scientific work, see R. Whitley, *The Intellectual and Social Organization of the Sciences* (Oxford: Clarendon Press, 1984).

7 C. Richet, "De l'influence des mouvements sur les idées," *Revue philosophique* 8 (1879):610–615; C. Richet, "Du somnambulisme provoqué," *Revue philosophique* 10 (1880):337–374, 462–493; H. Beaunis, "L'experiementation en psychologie par le somnambulisme provoqué," *Revue philosophique* 20 (1885):1–36; H. Beaunis, "Etudes physiologiques et psychologiques sur le somnambulisme provoqué" (Paris: Alcan, 1886); A. Binet and C. Féré, "Le hypnotisme chez les hysteriques: Le transfert psychique," *Revue philosophique* 19 (1885):1–25; J. Delboeuf, "La memoire chez les hypnotisés," *Revue philosophique* 21 (1886): 441–472; J. Delboeuf, "De l'influence de l'education et de l'imitation dans le somnambulisme provoqué," *Revue philosophique* 22 (1886):146–171. My account of this work was originally published as part of a paper on "The origins of the psychological experiment as a social institution," *American Psychologist* 40 (1985): 133–140.

8 The social structure of the clinical experiment was much closer to that of most contemporary psychological experiments than was the Leipzig experiment. In this connection it is interesting to note that the period of concern with the social psychology of the psychological experiment was ushered in by an analogy between role playing in hypnosis and role playing in experimental situations. See M. T. Orne, "On the social psychology of the psychological experiment: With particular reference to demand characteristics and their implications," *American Psychologist* 17 (1962):776–783; also M. T. Orne, "Hypnosis, motivation, and the ecological validity of the psychological experiment," *Nebraska Symposium on Motivation* 18 (1970):187–265.

9 A. Binet, "Recherches sur les mouvements chez quelques jeunes enfants," *Revue philosophique* 29 (1890):297–309.

10 *Grand Larousse de la langue française,* vol. 6 (Paris: Librairie Larousse, 1973); E. Littré, *Dictionnaire de la langue française,* vol. 7 (Paris: Gallimard-Hachette 1968).

11 E.g., in C. Féré, "Sensation et mouvement," *Revue philosophique* 20 (1885):337–368. This usage is also found in the titles of publications, e.g., by Bottey, "Des suggestions provoquées a l'état de veille chez les hysteriques et chez les sujet sains" (1884), and by Brémaud, "Hypnotisme chez des sujet sains" (1883), as cited in M. Dessoir, *Bibliographie des Hypnotismus* (Berlin: Duncker, 1888).

12 *Oxford English Dictionary,* vol. 10 (Oxford: Clarendon Press, 1933). The term crops up occasionally in James Braid's *Neurypnology* of 1843; see *Braid on Hypnotism: The Beginnings of Modern Hypnosis* (New York: Julian Press, 1960), p. 209. In later years psychic researchers sometimes investigated hypnotic phenomena and used the term, probably as part of an attempt to establish their scientific credentials. See C. S. Alvarado, "Note on the use of the term *Subject* in pre–1886 discussions of thought transference," *American Psychologist* 42 (1987):101–102.

13 G. S. Hall, "Reaction time and attention in the hypnotic state," *Mind* 8 (1883):170–182; E. Gurney, "The stages of hypnotism," *Mind* 9 (1884):110–121.

14 G. S. Hall and H. H. Donaldson, "Motor sensations on the skin," *Mind* 10 (1885):557–572; G. S. Hall and J. Jastrow, "Studies of rhythm," *Mind* 11 (1886): 55–62.

15 J. M. Cattell, "The time taken up by cerebral operations," *Mind* 11 (1886):220–242; J. M. Cattell and S. Bryant, "Mental association investigated by experiment," *Mind* 14 (1889):230–250.

16 See E. Gurney, "Further problems of hypnotism," *Mind* 12 (1887):112–222; E. Gurney, F. W. H. Myers, and F. Podmore, *Phantoms of the Living* (London: Trubner, 1886); J. Jacobs, "Experiments on 'prehension,' " *Mind* 12 (1887): 75–79.

17 The anthropometric laboratory subsequently reopened at the Science Museum, South Kensington. For an account of these laboratories and Galton's use of the knowledge they produced, see K. Pearson, *The Life, Letters and Labours of Francis Galton,* vol. 2 (Cambridge: Cambridge University Press, 1924), pp. 357–359; 370–386; also F. Galton, *Memories of My Life* (London: Methuen, 1908) pp. 244–250.

18 L. S. Hearnshaw, *A Short History of British Psychology 1840–1940* (London: Methuen, 1964), p. 19. D. W. Forrest, *Francis Galton: The Life and Work of a Victorian Genius* (London: Paul Elk, 1974), p. 37; K. Pearson, *The Life, Letters and Labours of Francis Galton,* vol. 1 (Cambridge: Cambridge University Press, 1914), p. 157. On the significance of phrenology for some of Galton's cohorts, see

R. M. Young, *Mind, Brain and Adaptation in the Nineteenth Century* (Cambridge: Cambridge University Press, 1970).

19 Galton was notoriously wrong about the general importance of the performances he measured – such matters as visual acuity and strength of hand grip – but this does not of course affect the significance of his research goals. Those who followed in his footsteps eventually came up with measures of less trivial performances; see R. Fancher, *The Intelligence Men* (New York: Norton, 1985).

20 R. Cooter, *The Cultural Meaning of Popular Science: Phrenology and the Organization of Consent in 19th Century Britain* (Cambridge: Cambridge University Press, 1984). On the relationship between the specialization of occupations and the emerging interest in the assessment of "individual differences," see also A. R. Buss, "Galton and the birth of differential psychology and eugenics: Social, political and economic forces," *Journal of the History of the Behavioral Sciences* (1976):12, 47–58.

21 A contractual model to describe the social situation in all types of psychological experiments has been proposed by H. Schuler, *Ethical Problems in Psychological Research* (New York: Academic Press, 1982). Although all experimental situations have their contractual aspects, the implication that all such situations are *essentially* contractual in nature is unfortunate. Such an approach may blind one to the possibilities of divergent types of experimental situations within which the contractual aspects may vary greatly in importance and significance. Moreover, these different types of experimental situations may entail different ethical considerations, and what is ethically appropriate in one context may not be so in another.

22 This terminology has been common since R. Rosenthal and R. L. Rosnow, eds., *Artifact in Behavioral Research* (New York: Academic Press, 1969). For a broader view on "artifacts," see R. M. Farr, "On the social significance of artifacts in experimenting," *British Journal of Social and Clinical Psychology* 17 (1978): 299–306.

23 These matters have been extensively explained in the work of C. Argyris; see, e.g., C. Argyris, "Some unintended consequences of rigorous research," *Psychological Bulletin* 70 (1969):185–197.

24 "Because of Charcot's paternalistic attitude and his despotic treatment of students, his staff never dared contradict him; they therefore showed him what they believed he wanted to see. After rehearsing the demonstrations, they showed the subjects to Charcot, who was careless enough to discuss their cases in the patients' presence. A peculiar atmosphere of mutual suggestion developed between Charcot, his collaborators, and his patients, which would certainly be worthy of an accurate sociological analysis." It would indeed. H. F. Ellenberger, *The Discovery of the Unconscious* (New York: Basic Books, 1970), p. 98.

25 For an analysis of the experiments by Titchener's students, written long after the dust had settled, see G. Humphrey, *Thinking: An Introduction to Its Experimental Psychology* (London: Methuen, 1950), pp. 119–129.

26 In the Leipzig mold: G. S. Hall and Y. Motora, "Dermal sensitiveness to gradual pressure changes," *American Journal of Psychology* 1 (1887):72–98; in the tradition of experimental hypnosis studies: G. S. Hall, "Reaction time." Those of Hall's studies that were closer to Galtonian practices will be discussed in the next chapter; they were the innovative ones.

27 Cattell contributed the term "mental test": J. M. Cattell, "Mental tests and measurements," *Mind* 15 (1890):373–381; see also J. M. Cattell and L. Farrand,

"Physical and mental measurements of the students of Columbia University," *Psychological Review* 3 (1896):618–648.

28 For a comprehensive overview, see J. A. Popplestone and M. White McPherson, "Pioneer psychology laboratories in clinical settings," in J. Brozek, ed., *Explanations in the History of Psychology in the United States* (Lewisburg, Penn.: Bucknell University Press, 1984), pp. 196–272.

29 E. L. Thorndike, *Animal Intelligence* (New York: Macmillan, 1911), p. 3.

30 For the period 1934–1936 I was only able to locate three American studies involving an exchange of experimenter and subject roles that were not in the area of sensation and perception, although in the previous decade the practice is still in wider use.

Chapter 5. The triumph of the aggregate

1 J. Ziman, *Public Knowledge: The Social Dimensions of Science* (Cambridge: Cambridge University Press, 1968).

2 S. Woolgar, "Discovery: Logic and sequence in a scientific text," and N. Gilbert and M. Mulkay, "Contexts of scientific discourse: Social accounting in experimental papers," both in K. D. Knorr, R. Krohn, and R. Whitley, eds., *The Social Process of Scientific Investigation* (Dordrecht: Reidel, 1980), pp. 239–268 and 269–294. See also R. Whitley, "Knowledge producers and knowledge acquirers," in T. Shinn and R. Whitley, eds., *Expository Science: Forms and Functions of Popularization* (Dordrecht: D. Reidel, 1985), pp. 3–28.

3 K. D. Knorr-Cetina, *The Manufacture of Knowledge* (New York: Pergamon, 1981).

4 This divergence is analyzed further in K. Danziger, "The positivist repudiation of Wundt," *Journal of the History of the Behavioral Sciences* 15 (1979):205–230.

5 In the case of Claude Bernard, probably the most celebrated exponent of this approach, advocacy of its virtues was coupled with derision for the statistical approach: "We shall therefore try to find these experimental, physiological conditions, instead of tabulating the variations in phenomena and taking averages as expressions of reality; we should thus reach conclusions based on correct statistics, but with no more scientific reality than if they were wholly arbitrary." Claude Bernard, *An Introduction to the Study of Experimental Medicine* (1865; repr. New York: Dover, 1957), pp. 116ff.

6 For Pavlov's investigative practice, see his *Conditioned Reflexes: An Investigation of the Physiological Activity of the Cerebral Cortex,* trans. and ed. G. V. Anrep (London: Oxford University Press, 1927).

7 An instructive example of this conception of psychological experimentation is provided by Titchener's defense of classical reaction-time studies against the criticism of Baldwin who preferred a typological interpretation of individual differences observed in such situations. See E. B. Titchener, "The type theory of simple reactions," *Mind* n.s. 4 (1895):506–514. The fundamental implications of the Baldwin–Titchener controversy have been noted more recently by D. L. Krantz, in *Schools of Psychology* (New York: Appleton Century Crofts, 1965), and by Gernot Böhme, "Cognitive norms, knowledge interests and the constitution of the scientific object: A case study in the functioning of rules for experimentation," in E. Mendelsohn, P. Weingart, and R. Whitley, eds., *The Social Production of Scientific Knowledge* (Dordrecht: Reidel, 1977). As representatives of major historical orientations within psychology, Baldwin and Titchener were not on the same level, however. Whereas Titchener gave highly coherent expression to a well-established position, then at the height of its development, Baldwin represented a new viewpoint whose methodological implications had not been fully worked out.

8 K. H. Metz, "Paupers and numbers: The statistical argument for social reform in Britain during the period of industrialization," in L. Krüger, L. J. Daston, and M. Heidelberger, eds., *The Probabilistic Revolution,* vol. 1: *Ideas in History* (Cambridge, Mass.: MIT Press, 1987), pp. 337–350; and T. M. Porter, *The Rise of Statistical Thinking 1820–1900* (Princeton: Princeton University Press, 1986), chap. 1.

9 For some of the historical background on questionnaires, see R. H. Gault, "A history of the questionnaire method of research in psychology," *Pedagogical Seminary* 14 (1907):366–383.

10 S. Jaeger, "Volkspsychologie, Statistik und Sozialreform," in S. Bem, H. Rappard, and W. van Hoorn, eds., *Studies in the History of Psychology and the Social Sciences,* vol. 2 (Leiden: Psychologisch Instituut Rijksuniversiteit Leiden, 1984), pp. 148–160.

11 G. S. Hall, "Contents of children's minds on entering school," *Pedagogical Seminary* 1 (1891):139–173.

12 B. Sigismund, *Kind und Welt* (Braunschweig: Vieweg, 1856). Nineteenth-century medical literature played an important role in establishing a place for statistics in scientific rhetoric. See J. H. Cassedy, *American Medicine and Scientific Thinking 1800–1860* (Cambridge, Mass.: Harvard University Press, 1986).

13 C. Darwin, *The Expression of the Emotions in Man and Animals* (1872; repr. Chicago: University of Chicago Press, 1965); F. Galton, *English Men of Science* (1874; repr. London: Frank Cass, 1970), and *Inquiries into Human Faculty* (London: Macmillan, 1883).

14 The pages of Hall's *Pedagogical Seminary* contain many studies of this type. For example, in the third volume of 1895–1896 we find: J. A. Jancock, "A preliminary study of motor ability in children" – 158 subjects; M. Schallenberger, "A study of children's rights as seen by themselves" – over 3,000 subjects; M. C. Holmes, "The fatigue of a school hour" – 150 subjects; E. Barnes, "Punishment as seen by children" – 4,000 subjects; K. Fackenthal, "The emotional life of children" – 200 subjects; M. A. Herrick, "Children's drawings" – 450 subjects; H. C. Kratz, "Characteristics of the best teacher as recognized by children" – 2,400 subjects; H. P. Kennedy, "Effect of high-school work upon girls during adolescence" – 129 subjects.

15 See Porter, *Statistical Thinking,* pp. 57–70. In his earliest statement on the methods of psychology, the introduction to the *Beiträge zur Theorie der Sinneswahrnehmung* (Leipzig and Heidelberg: C. F. Winter, 1862), Wundt mentions both statistics and experiment as the methods for placing psychology on a scientific footing. However, by the time his laboratory had begun operating in Leipzig, statistics had been very much downgraded compared to experiment. See, e.g., his *Methodenlehre* of 1883, p. 494 (W. Wundt, *Logik,* vol. 2 [Stuttgart: Enke, 1883]).

16 L. A. Quetelet, *Sur l'homme et le développement de ses facultés* (Paris: Bachelier, 1835); Eng. trans., *A Treatise on Man and the Development of His Faculties* (Gainesville, Fla.: Scholars' Facsimiles, 1969).

17 S. P. Turner, *The Search for a Methodology of Social Science* (Dordrecht: Reidel, 1986), chap. 4. See also G. Gigerenzer et al., *The Empire of Chance* (Cambridge: Cambridge University Press, 1989), chap. 2.

18 Porter, *Statistical Thinking,* chap. 6; and the related discussion in I. Hacking,

"Prussian numbers 1860–1882," in L. Krüger, L. J. Daston, and M. Heidelberger, eds., *The Probabilistic Revolution,* vol. 1: *Ideas in History* (Cambridge, Mass.: MIT Press, 1987), pp. 377–394.

19 H. T. Buckle, *History of Civilization in England,* 2 vols. (London: J. W. Parker, 1858–1861).

20 On the fundamental distinction between "essentialism" and "population think-ing" in biology, see E. Mayr, *The Growth of Biological Thought: Diversity, Evo-lution and Inheritance* (Cambridge, Mass.: Harvard University Press, 1982).

21 On the implications of the Galtonian basis for psychology, and especially for its role in society, see N. Rose, *The Psychological Complex: Psychology, Politics and Society in England 1869–1939* (London: Routledge & Kegan Paul, 1985).

22 The distinction was quite clear to some nineteenth-century statisticians. F. Y. Edgeworth, for instance, expressed it in terms of a distinction between "obser-vations" and "statistics": "Observations and statistics agree in being grouped about a Mean; they differ, in that the Mean of observations is real, of statistics is fictitious. The mean of observations is a cause . . . the mean of statistics is a description, a representative quantity put for a whole group. . . . Different measurements of the same man are observations; but measurements of different men, grouped round l'homme moyen, are prima facie at least statistics." Edgeworth, 1885, cited in S. M. Stigler, *The History of Statistics: The Measurement of Uncertainty before 1900* (Cambridge, Mass.: Harvard University Press, 1986), p. 309.

23 One still meets this approach in the well-known twentieth-century work of Hall's student, Arnold Gesell; see, e.g., A. Gesell et al., *First Five Years of Life: A Guide to the Study of the Preschool Child* (New York: Harper, 1940); or A. Gesell et al., *The Child from Five to Ten* (New York: Harper, 1946).

24 Explicit comments on the shortcomings of the "statistical" approach to psy-chological questions were common in the European literature – e.g., Binet in *L'étude experimentale de l'intelligence* (Paris: Schleicher, 1903): "The Americans who love to do things big, often publish experiments made on hundreds and even thousands of persons. They believe that the conclusive value of a work is propor-tional to the number of observations made. This is an illusion. . . . If I have been able to throw some light by the attentive study of two subjects, it is because I have seen their behavior from day to day and have probed it over a period of several years. . . . We should prefer experiments that we can make on persons whose char-acter and way of life are familiar to us" (Eng. trans. from T. H. Wolf, *Alfred Binet* [Chicago: University of Chicago Press, 1973], p. 120). Some samples of German criticisms of large samples will be found in H. Münsterberg, *Grundzüge der Psy-chologie* (Leipzig: Barth, 1900); C. Stumpf, "Zur Methodik der Kinderpsycholo-gie," *Zeitschrift für pädagogische Psychologie und Pathologie* 2 (1900):17–21; W. Ament, "Logik der statistischen Methode," introduction to *Die Entwicklung der Pflanzenkenntnis beim Kinde und bei Völkern* (Berlin: Reuther and Reichard, 1904), pp. 3–14. The last-named author sums up a lengthy discussion with the epigrammatic statement (p. 14), "Die Masse aber trübt dem Forscher den Blick" (the mass blurs the investigator's gaze).

25 An uncritical acceptance of "official" categories like crime, suicide, etc., had of course been the basis for the application of statistics to social phenomena, starting with Quetelet. Now, the same naive acceptance of conventional categories was transferred to psychology. In both cases the investigators were able to reap the immediate advantages of using the same conceptual network as the one in which ordinary people organized their experience and expressed their concerns. The apparent practical relevance of such studies seemed obvious, whereas psychological experiments deconstructed common experience, often without reestablishing links with it. Today, there is much greater awareness of the problematic nature of the social categories that form the basis both of social statistics and of experiments on social conduct; e.g., J. Irvine, I. Miles, and J. Evans, eds., *Demystifying Social Statistics* (London: Pluto, 1979); and K. J. Gergen, *Toward Transformation in Social Knowledge* (New York: Springer, 1982).

26 "The school is itself a kind of intelligence test, and the eventual modernization of American education meant the intensifying of actions that could produce estimates of learning ability." D. Calhoun, *The Intelligence of a People* (Princeton: Princeton University Press, 1973), p. 76. See also Rose, *Psychological Complex.*

27 Even William James was not immune to the appeal of such a science, although it was left to others to convert his ideas into appropriate practices: "All natural sciences aim at practical prediction and control and in none of them is this more the case than in psychology to-day. We live surrounded by an enormous body of persons who are most definitely interested in the control of states of mind, and in incessantly craving for a sort of psychological science which will teach them how to *act*. What every educator, every asylum superintendent, asks of psychology is practical rules. Such men care little or nothing about the ultimate philosophic grounds of mental phenomena, but they do care immensely about improving the ideas, dispositions, and conduct of the particular individuals in their charge." W. James, "A plea for psychology as a 'natural science,'" *Philosophical Review* 1 (1892):146–153 (p. 148). The premise, "that individuals in the national population were best described, analyzed, and interpreted by reference to the group in the national population to which they 'belonged,' " has been called "the fundamental methodological premise of American evolutionary natural and social science." See H. Cravens, "Applied science and public policy," in M. Sokal (ed.), *Psychological Testing and American Society* (New Brunswick, N.J.: Rutgers University Press, 1987), pp. 158–194 (p. 179).

28 Most of the prominent figures in the early years of modern American psychology had little respect for the tradition of painstaking laboratory work represented by figures like Fechner, Wundt, and Titchener, either because of its "microscopic" character (James) or because of its purely academic interest (Hall). Also, this kind of psychology contributed little or nothing to satisfying the broad social demands that the American tradition placed on psychology as a teaching subject; cf. F. C. French, "The place of experimental psychology in the undergraduate courses," *Psychological Review* 5 (1898):510–512. The problems faced by esoteric laboratory work in a culture hostile to intellectual elitism were not unique to psychology, but also arose in other sciences. See G. H. Daniels, *Science in American History: A Social History* (New York: Knopf, 1971), pp. 281ff.

29 In the original coding five categories of data were distinguished: (a) studies based exclusively on individual data; (b) studies essentially based on individual data but including some summaries; (c) studies essentially based on group data but including occasional reference to individual cases for purposes of illustration; (d) studies based exclusively on group data; and (e) truly mixed studies where both kinds of data carried some weight. However, categories (b) and (c) were found to show the same pattern of variation as the extreme categories (a) and (d). The pair (a) and (b) and the pair (c) and (d) were therefore collapsed into two combined categories. Together with (e) this resulted in a threefold classification which represents the overall trend more economically. Agreement among raters on the inclusion of articles in these categories ranged from 91.6 percent for the overall "group data" category to 83.5 percent for the overall "individual data" category.

30 In the case of two of the journals the first period actually represented the late 1910s because they only began publication in 1916 or 1917, and because they missed a year or two due to the war.

31 Some pertinent considerations are to be found in the following: W. van Hoorn, "Psychology and the reign of technology," in S. Bem, H. Rappard, and W. van Hoorn, eds., *Studies in the History of Psychology and the Social Sciences*, vol. 1 (Leiden: Psychologisch Instituut, 1983), pp. 105–118; W. van Hoorn and T. Verhave, "Socio-economic factors and the roots of American psychology: 1865–1914," *Annals of the New York Academy of Sciences* 291 (1977):203–221.

32 In this context the term "laboratory" has a social rather than a physical definition; it is not constituted by the presence of apparatus but by an investigative situation.

33 The importance of World War I intelligence testing for the social impact of psychology as a discipline is well recognized. See F. Samelson, "World War I intelligence testing and the development of psychology," *Journal of the History of the Behavioral Sciences* 13 (1977):274–282; and F. Samelson, "Putting psychology on the map: Ideology and intelligence testing," in A. R. Buss, ed., *Psychology in Social Context* (New York: Irvington, 1979), pp. 103–168.

34 See D. Bakan, *On Method* (San Francisco: Jossey Bass, 1967), chap. 1.

35 The metaphorical use of "organism" in connection with statistically constituted entities goes back to Gustav Theodor Fechner. See his *Kollektivmasslehre* (Leipzig: Engelmann, 1897).

36 At this stage in the history of psychology such transfers were often made in the service of one or another ideological position. A very effective means of promoting the social standing of psychology as a discipline was to hitch its investigative practices to one or another preexisting social ideology. Thus, intelligence testers claimed to have found the measure that radical hereditarians needed to enhance the plausibility of their eugenicist program, and behaviorists claimed to have found the techniques on which the effectiveness of programs of social control depended. The fundamental features that these two positions had in common were probably far more important than their superficial differences. It has often been pointed out that they shared the same undemocratic goals – e.g., J. Harwood, "The IQ in History," *Social Studies of Science* 13 (1983):465–477; R. Marks, *The Idea of IQ* (Washington, D.C.: University Press of America, 1981); F. Samelson, "On the science and politics of the IQ," *Social Research* 42 (1975):467–488. They also shared a simplistic intellectual style that was inseparable from their very real political role in mobilizing support for broad social programs. The intimate involvement with such ideological positions produced a simplistic formulation of psychological issues that left its traces on the discipline for a long time to come.

Chapter 6. Identifying the subject in psychological research

1 For a comprehensive account of the shaping of this double thrust, see J. M. O'Donnell, *The Origins of Behaviorism: American Psychology, 1870–1920* (New York: New York University Press, 1985).

2 By 1938, the proportion of published American psychological research studies that were based on work done with animal subjects had reached 15 percent of the total. This proportion had increased steadily over four decades. J. S. Bruner and G. W. Allport, "Fifty years of change in American psychology," *Psychological Bulletin* 37 (1940):757–776 (p. 764). Moreover, those who worked with animal subjects tended to show a higher research productivity than those who worked with human subjects – a likely index of high status in the discipline. S. W. Fernberger, "The scientific interests and scientific publications of the members of the American Psychological Association," *Psychological Bulletin* 35 (1938):261–281. A quarter of the papers delivered at the 1939 annual meeting of the APA were based on animal research; see Bruner and Allport, ibid. p. 765.

3 The problem is not peculiar to scientific psychology; see K. D. Knorr-Cetina, "The ethnographic study of scientific work: Towards a constructivist interpretation of science," in K. D. Knorr-Cetina and M. Mulkay, eds., *Science Observed: Perspectives on the Social Study of Science* (London and Beverly Hills, Calif.: Sage, 1983), pp. 115–140.

4 See Popplestone and McPherson, "Pioneer psychology laboratories."

5 This feature is discussed in chapter 9.

6 As indicated in the previous chapter, this approach was particularly common in the area of child study, but it was not limited to this area. Early examples of psychological census taking among adults are: J. Jastrow, "Community and association of ideas: A statistical study," *Psychological Review* 1 (1894):152–158; and M. W. Calkins, "Synaesthesia," *American Journal of Psychology* 7 (1895):90–107.

7 See, e.g., R. Smart, "Subject selection bias in psychological research," *Canadian Psychologist* 7 (1966):115–121; D. P. Schultz, "The human subject in psychological research," *Psychological Bulletin* 72 (1969):214–228; K. L. Higbee and M. G. Wells, "Some research trends in social psychology during the 1960's," *American Psychologist* 27 (1972):963–966.

8 J. F. Dashiell, "Note on use of the term 'observer,' " *Psychological Review* 36 (1929):550.

9 M. Bentley, " 'Observer' and 'subject,' " *American Journal of Psychology* 41 (1929):682.

10 J. F. Dashiell, "A reply to Professor Bentley," *Psychological Review* 37 (1930):183–185.

11 The empirical studies published in the *Psychological Review* before it became a theoretical journal tended to have the same character as those published in the *American Journal of Psychology,* and later, in the *Journal of Experimental Psychology. Psychological Monographs,* however, showed considerable variation on the kinds of studies published, also from one time period to another.

12 This is particularly the case for the *Journal of Educational Psychology.*

Chapter 7. Marketable methods

1 A telling little interchange is recalled by Charles Judd who was to achieve a position of considerable prominence in American educational psychology. In 1898, shortly after his return from Wundt's laboratory, he had been appointed to a professorship at the School of Pedagogy of New York University. The memory of the following incident seems to have remained with him ever after: "I recall very well that I had on one occasion been lecturing enthusiastically on Weber's Law to a class of New York City teachers who were seeking increases in their salaries by listening to me, when I was interrupted by one of my gray-haired auditors with this question: 'Professor, will you tell us how we can use this principle to improve our teaching of children?' I remember that question better than I do my answer." C. H. Judd in C. Murchison, ed., *A History of Psychology in Autobiography,* vol. 2 (Worcester, Mass: Clark University Press, 1932), pp. 221–222.

2 L. Zenderland, "Education, evangelism, and the origins of clinical psychology: The child-study legacy," _Journal of the History of the Behavioral Sciences_ 24 (1988):152–165.

3 The rate of expansion of the American educational system had reached staggering proportions. Between 1890 and 1920 there was at least one public high school built for each day of the calendar year and the enrollment increase was of the order of 1,000 percent. Between 1902 and 1913 public expenditure on education more than doubled; between 1913 and 1922 it tripled. The relation between changes in American education and the requirements of the social system has often been explored; see, e.g., C. J. Karrier, ed., _Shaping the American Educational State_ (New York: Free Press, 1975); S. Bowles and H. Gintis, _Schooling in Capitalist America_ (New York: Basic Books, 1976); J. Spring, _Education and the Rise of the Corporate State_ (Boston: Beacon Press, 1972).

4 A. G. Powell, "Speculations on the early impact of schools of education on educational psychology," _History of Education Quarterly_ 11 (1971):406–412, who speaks of "the deeply felt need of careerist male educators to separate themselves as much as possible from classroom teaching. Teaching, by and large, was viewed as transient, unrewarding, unprofessional, and female. Separate organizations, journals, and university courses were demanded by former teachers seeking a better and more respectable career identity through specialized administrative positions" (p. 410). See also R. E. Callahan and H. W. Button, "Historical change and the role of the man in the organization: 1865–1950," in D. E. Griffiths, ed., "Behavioral Science and Educational Administration," _Yearbook, National Society for the Study of Education_ 63 (1964):73–92. On the role of testing in the educational system of the time, see D. P. Resnick, "Educational policy and the applied historian," _Journal of Social History_ 14 (1981):539–559.

5 F. Bobbitt, "Some general principles of management applied to the problems of city-school systems," _Yearbook, National Society for the Study of Education_ 12 (1913):7–96 (p. 12).

6 For a sampling of such analogies, see R. E. Callahan, _The Cult of Efficiency_ (Chicago: University of Chicago Press, 1962); and D. Tyack and E. Hansot, _Managers of Virtue: Public School Leadership in America, 1820–1980_ (New York: Basic Books, 1982).

7 F. E. Spaulding, "The application of the principles of scientific management," (NEA _Proceedings,_ 1913) as cited in Callahan, _Cult of Efficiency,_ p. 68.

8 In the standard text on school administration, the role of research departments in school districts is described as follows: "to study the needs of life and the industries, with a view to restating the specifications for the manufacture of the educational output; to study means for increasing the rate of production, and for eliminating the large present waste in manufacture; . . . to test out different methods of procedure, and gradually to eliminate those which do not give good results." E. P. Cubberley, _Public School Administration_ (Boston: Houghton-Mifflin, 1916), p. 336. See also G. D. Strayer, "Measuring results in education," _Journal of Educational Psychology_ 2 (1911):3–10.

9 See G. Joncich, _The Sane Positivist: A Biography of Edward L. Thorndike_ (Middletown, Conn.: Wesleyan University Press, 1968). On the liaison between American education and psychological research, see also E. V. Johanningmeier, "American educational research: Applications and misapplications of psychology to education," in T. V. Smith and D. Hamilton, eds., _The Meritocratic Intellect: Studies in the History of Educational Research_ (Aberdeen: Aberdeen University Press, 1980), pp. 41–57.

10 E. C. Sanford, "Experimental pedagogy and experimental psychology," *Journal of Educational Psychology* 1 (1910):590–595. See also C. Seashore, "The educational efficiency engineer," *Journal of Educational Psychology* 4 (1913):244. For other psychologists involved in this area, see R. I. Watson, "A brief history of educational psychology," *The Psychological Record* 11 (1961):209–242.

11 R. L. Church, "Educational psychology and social reform in the progressive era," *History of Education Quarterly* 11 (1971):390–405.

12 On the significance of the war work for the establishment of psychology as a socially relevant science, see F. Samelson, "Putting psychology on the map," pp. 103–168.

13 R. S. Woodworth, *Psychological Issues* (1920; repr. New York: Columbia University Press, 1939), p. 385. In an autobiographical statement, J. Jastrow, the first psychologist who was never anything else but a psychologist, writes: "To speak of the renaissance of psychology, especially in the American setting, without explicit recognition of the practical motive would be a glaring omission; for that renaissance found its momentum in the appeal to psychology for the regulation of human affairs. Applied psychology is in many a quarter the pay-vein that supports the mine. The educational application is the oldest and most comprehensive." C. Murchison, ed., *A History of Psychology in Autobiography,* vol. 1 (Worcester, Mass.: Clark University Press, 1930), p. 155. By 1910, three-quarters of those American psychologists who indicated an active interest in psychological applications expressed that interest in the field of education; see J. M. O'Donnell, *The Origins of Behaviorism: American Psychology, 1870–1920* (New York: New York University Press, 1985), p. 227.

14 There is an extensive literature in this area. See, e.g., L. Baritz, *The Servants of Power* (Middletown, Conn.: Wesleyan University Press, 1965); D. S. Napoli, *Architects of Adjustment: The History of the Psychological Profession in the United States* (Port Washington, N.Y.: Kennikat Press, 1981); R. Marks, *The Idea of IQ* (Washington, D.C.: University Press of America, 1981); J. Harwood, "The IQ in history," *Social Studies of Science* 13 (1983):465–477; M. M. Sokal, ed., *Psychological Testing and American Society* (New Brunswick, N.J.: Rutgers University Press, 1987). On the British case, see G. Sutherland, *Ability, Merit and Measurement* (Oxford: Clarendon Press, 1984).

15 E. L. Thorndike, "The contribution of psychology to education," *Journal of Educational Psychology* 1 (1910):5–12.

16 E. C. Sanford, "Methods of research in education," *Journal of Educational Psychology* 3 (1912):303–315.

17 L. J. Cronbach, "The two disciplines of scientific psychology," *American Psychologist* 12 (1957):671–684. Between 1921 and 1936 some five thousand articles on mental testing were published in various journals; see E. B. South, *An Index of Periodical Literature on Testing, 1921–1936* (New York: The Psychological Corporation, 1937).

18 See, e.g., A. Ciocco, "The background of the modern study of constitution," *Bulletin of the History of Medicine* 4 (1936):23–28; J. Graham, "Lavater's physiognomy in England," *Journal of the History of Ideas* 22 (1961):561–572.

19 See S. J. Gould, *The Mismeasure of Man* (New York: Norton, 1981).

20 A. Binet and V. Henri, "La psychologie individuelle," *L'Anée psychologique* 2 (1895):411–465. See also S. E. Sharp, "Individual psychology: A study in psychological method," *American Journal of Psychology* 10 (1899):329–391; and T. H. Wolf, *Alfred Binet* (Chicago: University of Chicago Press, 1973).

21 W. Stern, *Die differentielle Psychologie in ihren methodischen Grundlagen* (Leipzig: Barth, 1911). Stern had published his first review of the literature on individual differences in 1900, several years before Galtonian mental tests came to dominate the field. His final position is represented by his *General Psychology from the Personalistic Standpoint* (New York: Macmillan, 1938). For a sympathetic American voice, see M. Bentley, "Individual psychology and psychological varieties," *American Journal of Psychology* 52 (1939):300–301.

22 J. M. Baldwin, "Types of reaction," *Psychological Review* 2 (1895):259–273; and "The type theory of reactions," *Mind* n.s. 5 (1896):81–90.

23 This assumption was usually expressed in the form of Thorndike's axiom that all qualitative differences between individuals are reducible to quantitative differences: "All intelligible differences are ultimately quantitative. The difference between any two individuals, if describable at all, is described by comparing the amounts which A possesses of various traits with amounts which B possesses of the same traits. In intellect and character, differences of kind between one individual and another turn out to be definable, if defined at all, as compound differences of degree." E. L. Thorndike, *Individuality* (Boston: Houghton-Mifflin, 1911), p. 5.

24 J. Valsiner, ed., *The Individual Subject and Scientific Psychology* (New York: Plenum, 1986).

25 For more extensive treatment of this topic, see N. Rose, *The Psychological Complex: Psychology, Politics and Society in England 1869–1939* (London: Routledge and Kegan Paul, 1985): "This psychology sought to establish itself by claiming its ability to deal with the problems posed for social apparatuses by dysfunctional conduct. . . . psychological normality was conceived of as . . . an absence of social inefficiency: that which did not need to be regulated" (pp. 5–6).

26 See R. Cooter, *The Cultural Meaning of Popular Science: Phrenology and the Organization of Consent in 19th Century Britain* (Cambridge: Cambridge University Press, 1984); and D. Bakan, "The influence of phrenology on American psychology," *Journal of the History of the Behavioral Sciences* 2 (1966):200–220.

27 The comprehensive history of the emergence of examinations as a major social institution in the nineteenth century remains to be written. Existing sources are widely scattered. The following provide access to material in this area: Rose, *Psychological Complex;* J. Roach, *Public-Examinations in England 1850–1900* (Cambridge: Cambridge University Press, 1971); R. J. Montgomery, *Examinations: An Account of their Evolution as Administrative Devices in England* (London: Longmans, 1965). Galton appears to have been fascinated by examinations and used results from the Indian Civil Service examinations in his attempts at developing a measure of statistical correlation. See R. E. Fancher, "The examined life: Competitive examinations in the thought of Francis Galton," *History of Psychology Newsletter* 16 (1984):13–20.

28 A contemporary reviewer presented the difference in practice as follows: "The French psychologists have tended on the whole to study the individual's reactions with a view to understanding the total adjustment of the organism or personality. The English and Americans, in large part, have concerned themselves with expressing the reactions in statistical terms. . . . Moreover, those who view intelligence testing merely as an aid to clinical diagnosis tend toward the older view, the French, while those who deal in mass materials would emphasize the latter and tend to define intelligence in statistical formulae"; see K. Young, "The history of mental testing," *The Pedagogical Seminary* 31 (1923):1–48 (p. 45). See also R. E. Fancher, *The Intelligence Men* (New York: Norton, 1985).

29 J. M. O'Donnell, "The clinical psychology of Lightner Witmer," *Journal of the History of the Behavioral Sciences* 15 (1979):3–17; B. Richards, "Lightner Witmer and the project of psychotechnology," *History of the Human Sciences* 1 (1988):201–219; J. A. Popplestone and M. W. McPherson, "Pioneer psychology laboratories in clinical settings," in J. Brozek, ed., *Explorations in the History of Psychology in the United States* (Lewisburg, Penn.: Bucknell University Press, 1984), pp. 196–272.

30 See F. Samelson, "Was early mental testing (a) racist inspired, (b) objective science, (c) a technology for democracy, (d) the origin of multiple-choice exams, (e) none of the above?" in Sokal, *Psychological Testing and American Society,* pp. 113–127.

31 Statisticians who did not share the Pearsonian world view were critical of what they considered a misuse of their craft by the mental-test movement. In his review of an early text on the statistics of mental testing, the British statistician Yule wrote: "Statistical methods...should be regarded as ancillary, not essential. They are essential where the subject of investigation is itself an aggregate, as a swarm of atoms or a crowd. But here the subject is the individual"; see G. Udny Yule, "Critical notice of Brown and Thomson's Essentials of Mental Measurement," *British Journal of Psychology* 12 (1921):105. On the fundamental differences between Yule and Pearson, see D. A. Mackenzie, *Statistics in Britain 1865–1930: The Social Construction of Scientific Knowledge* (Edinburgh: Edinburgh University Press, 1981).

32 R. S. Cowan, "Francis Galton's statistical ideas: The influence of eugenics," *Isis* 63 (1972):509–528; D. A. Mackenzie, *Statistics in Britain.*

33 It is more accurate to speak of a Galton–Pearson school, because while Galton provided the inspiration, Pearson provided the philosophically and technically complete formulation. However, the term "Galtonian" is more convenient as an adjectival form as long as it is understood as a kind of historical shorthand.

34 I previously presented this analysis in K. Danziger, "Statistical method and the historical development of research practice in American psychology," in G. Gigerenzer, L. Kruger, and M. Morgan, eds., *The Probabilistic Revolution: Ideas in Modern Science,* vol. 2 (Cambridge, Mass.: MIT Press, 1987), pp. 35–47.

35 B. T. Baldwin et al., *A Survey of Psychological Investigations with Reference to Differentiations between Psychological Experiments and Mental Tests* (Swarthmore, Penn.: APA Committee on the Academic Status of Psychology, 1916).

36 The problem is discussed at length in J. T. Lamiell, *The Psychology of Personality: An Epistemological Inquiry* (New York: Columbia University Press, 1987). See also K. Danziger, "The methodological imperative in psychology," *Philosophy of the Social Sciences* 15 (1985):1–13.

37 See H. Cravens, *The Triumph of Evolution: American Scientists and the Heredity–Environment Controversy 1900–1941* (Philadelphia: University of Pennsylvania Press, 1978), and S. J. Gould, *The Mismeasure of Man* (New York: Norton, 1981).

38 The use of treatment groups for purposes of experimental control was virtually unknown in experimental psychology prior to the explosion of educational research using such experimental design; see R. L. Solomon, "An extension of control group design," *Psychological Bulletin* 46 (1949):132–150. Even investigators like Thorndike were slow to appreciate the value of this approach. As Boring has pointed out, in their classical study of transfer of training, E. L. Thorndike and R. S. Woodworth, "The influence of improvement in one mental function upon the efficiency of other functions," *Psychological Review* 8 (1901):247–261, the authors used a control group only in a subsidiary experiment to which they devoted less than 1 percent of their article. Nor did the situation change during the next fifteen years. See E. G. Boring, "The nature and history of experimental control," *American Journal of Psychology* 67 (1954):573–589.

39 Crude statistical comparison of groups subjected to different conditions or treatments was well known in nineteenth-century medicine, but there is no evidence that this had any influence on psychological practice. For the medical background, see A. M. Lilienfeld, "Ceteris Paribus: The evolution of the clinical trial," *Bulletin of the History of Medicine* 56 (1982):1–18.

40 R. H. Winch, "The transfer of improvement of memory in schoolchildren," *British Journal of Psychology* 2 (1908):284–293; Winch, "Some measurements of mental fatigue in adolescent pupils in evening schools," *Journal of Educational Psychology* 1 (1910):13–24 and 83–100; Winch, "Accuracy in 'transfer'?" *Journal of Educational Psychology* 1 (1910):557–589; Winch, "Mental fatigue in day school children, as measured by arithmetical reasoning," *British Journal of Psychology* 4 (1911):315–341; Winch, "Mental fatigue in day school children as measured by immediate memory," *Journal of Educational Psychology* 3 (1912):18–28 and 75–82; Winch, "Mental adaptation during the school day as measured by arithmetical reasoning," *Journal of Educational Psychology* 4 (1913):17–28 and 71–84.

41 Winch seems to have been aware of his status as a pioneer. In a section entitled "The scientific interest of the work" he states: "I have, I believe for the first time, employed the method of *equal groups* to the solution of questions of fatigue" (Winch, *Journal of Educational Psychology* 1[1910]:16; emphasis in original). He was also given due credit for what soon came to be known as the "method of equivalent groups."

42 J. P. Gilbert, "An experiment on methods of teaching zoology," *Journal of Educational Psychology* 1 (1910):321–332; M. E. Lakeman, "The whole and part method of memorizing poetry and prose," *Journal of Educational Psychology* 4 (1913):189–198; L. W. Kline, "Some experimental evidence in regard to formal discipline," *Journal of Educational Psychology* 5 (1914):259–266; G. C. Myers, "Learning against time," *Journal of Educational Psychology* 6 (1915):115–116; T. E. Mayman, "An experimental investigation of the book, lecture and experimental methods of teaching physics in elementary schools," *Journal of Educational Psychology* (1915):6, 246–250. Peter Shermer (personal communication) has pointed out that the paper by Kline probably contains the first use of the term "control group" in the psychological literature.

43 E. L. Thorndike, "Fatigue in a complex function," *Psychological Review* 21 (1914):402; H. E. Conrad and G. F. Arps, "An experimental study of economical learning," *American Journal of Psychology* 27 (1916):507–529.

44 G. Melcher, "Suggestions for experimental work," *Yearbook, National Society for the Study of Education,* 17, pt. 2 (1918):139–151. See also R. R. Rusk, *Experimental Education* (London and New York: Longmans, Green, 1919), pp. 10–11.

45 W. A. McCall, *How to Experiment in Education* (New York: Macmillan, 1923).

46 Ibid., p. 3.

47 R. A. Fisher, *Statistical Methods for Research Workers* (London: Oliver & Boyd, 1925). McCall's terminology was of course different from that of Fisher, but some of the techniques advocated were similar. The memory of McCall's text was invoked by D. T. Campbell and J. C. Stanley in their enormously influential *Experimental* *and Quasi-Experimental Designs for Research* (Chicago: Rand McNally, 1963). Randomization, another component of the new literature on experimental design, had a different background. See I. Hacking, "Telepathy: Origins of randomization in experimental design," *Isis* 79 (1988):427–451.

48 R. M. W. Travers, *How Research Has Changed American Schools* (Kalamazoo, Mich.: Mythos, 1983), pp. 154–157, 526.

49 For a detailed analysis of the paradigm case for experimenting in an industrial setting, see R. Gillespie, "The Hawthorne experiments and the politics of experimentation," in J. G. Morawski, ed., *The Rise of Experimentation in American Psychology* (New Haven, Conn.: Yale University Press, 1988), pp. 114–137.

50 Although laboratory investigators were slow to adopt treatment group methodology, the method became steadily more popular in the academic research literature. Use of treatment groups in the *Journal of Experimental Psychology* increased from only 6 percent of empirical studies in 1916–1920 to 15 percent in 1924–1926 and then to 23 percent in 1934–1936. The corresponding figures for the more conservative *American Journal of Psychology* were 2 percent (1914–1916), 8 percent (1924–1926), and 13 percent (1934–1936).

51 The idea of general laws of behavior that would be applied to social problems was a favorite theme of behaviorism. According to its founder, psychology should "establish laws or principles for the control of human action so that it can aid organized society in its endeavor to prevent failures in such adjustments." J. B. Watson, "An attempted formulation of the scope of behavior psychology," *Psychological Review* 24 (1917):329–352. But this perspective was by no means limited to behaviorists: "And it is the outstanding feature of our reconstructed psychology that it realized and accepted the obligation to apply to education, to social relations, to practical affairs, to the control of human behavior generally, normal and abnormal, desirable and undesirable, the conclusions arising from the scientific study of the mental side of man." See J. Jastrow, "The reconstruction of psychology," *Psychological Review* 34 (1927):169–195 (p. 170).

52 An occasional voice pointed out that the emperor's new statistical clothes could not hide his psychological nakedness: "A large part of the quantitative and statistical methodology now to be found in practical fields is associated rather with its use by certain psychologists and among biometricians than derived through a logical connection with psychology itself." Madison Bentley in C. Murchison, ed., *A History of Psychology in Autobiography,* vol. 3 (Worcester, Mass.: Clark University Press, 1936) p. 64. However, it is clear that Bentley was, and perceived himself as being, out of step with most of his fellow psychologists. The general issue is taken up again in chapter 9 of the present volume.

53 Part of the strong attraction that statistical theories of experimental design held for a later generation of experimenters was probably due to the possibilities they offered for transforming problems of individual history into statistical problems.

Chapter 8. Investigative practice as a professional project

1 The term "professional project" functions as a convenient shorthand for a complex of common interests and strategies for pursuing them. An extensive discussion of this complex is provided in M. S. Larson, *The Rise of Professionalism: A Sociological Analysis* (Berkeley: University of California Press, 1977). On the emergence of a professional identity among American psychologists, see T. M. Camfield, "The professionalization of American psychology," *Journal of the History of the Behavioral Sciences* 9 (1973):66–75.

2 Such concerns were far from being unique to psychologists. See R. Whitley, *The Intellectual and Social Organization of the Sciences* (Oxford: Clarendon Press, 1984).

3 On the "rhetorical fabric" that enveloped modern American psychology from its earliest years, see D. E. Leary, "Telling likely stories: The rhetoric of the new psychology, 1880–1920," *Journal of the History of the Behavioral Sciences* 23 (1987):315–331.

4 It needs emphasizing that the essential conflict was systemic and not a conflict of persons. Individuals might therefore express a mixture of interests, depending on specific circumstances. The resulting tensions could be displayed on an intra-personal as well as an interpersonal level. But they were an inherent feature of the psychological enterprise in this particular historical context. Relevant material is to be found in J. M. O'Donnell, "The crisis of experimentalism in the 1920s," *American Psychologist* 34 (1979):289–295; M. Sokal, "James McKeen Cattell and American Psychology in the 1920s," in J. Brozek, ed., *Explorations in the History of American Psychology* (Lewisburg, Penn.: Bucknell University Press, 1982), pp. 273–323; and D. S. Napoli, *Architects of Adjustment: The History of the Psychological Profession in the United States* (Port Washington, N.Y.: Kennikat Press, 1981).

5 The organizational ups and downs of this relationship have been described in D. S. Napoli, *Architects of Adjustment;* also in J. E. W. Wallin, "History of the struggles within the American Psychological Association to attain membership requirements, test standardization, certification of psychological practitioners, and professionalization," *Journal of General Psychology* 63 (1960):287–308; and partly in S. W. Fernberger, "The American Psychological Association: A historical summary, 1892–1930," *Psychological Bulletin* 29 (1932):1–89. See also E. R. Hilgard, *Psychology in America: A Historical Survey* (San Diego: Harcourt Brace Jovanovich, 1987) chap. 20.

6 The notion that technological change was due to the application of science was part of the popular rhetoric of science at the time, and psychologists were able to deploy it effectively because it was such a pervasive illusion. See M. H. Rose, "Science as an idiom in the domain of technology," *Science and Technology Studies* 5 (1987):3–11; and also the relevant parts of D. Bakan, "Politics and American psychology," in R. W. Rieber and K. Salzinger, eds., *Psychology: Theoretical–Historical Perspectives* (New York: Academic Press, 1980), pp. 125–144.

7 "Every psychologist knows how difficult it is to get appropriations and maintenance for purely scientific work, and how much more impressive to the powers that control money is something which is 'practical,' however flimsy and evanescent its 'practicality.' The amount of money wasted in practical work which might be saved if more were available for the fundamental work on which eventual practical applications depend, is, of course, enormous, and even in psychology it is relatively large." See K. Dunlap, "The experimental methods of psychology," in M. Bentley et al., *Psychologies of 1925* (Worcester, Mass.: Clark University, 1926) p. 342. Note how the author deploys the model of technology as applied science to defend the priority of basic research on purely economic grounds.

8 On the role of experimentalism as an emblem of high status within the hierarchy of the sciences, see Whitley, *Intellectual and Social Organization,* pp. 276–289. One institutional mechanism through which such a hierarchy of prestige maintained itself was to be found in self-selected national scientific bodies. During the period of primary interest here, only one (token?) nonexperimentalist (Terman) was elected to the National Academy of Sciences, as compared with seven other psychologists who were able to rely on their achievements in the laboratory (Angell, Dodge, Pillsbury, Seashore, Stratton, Woodworth, Yerkes).

9 It should be remembered that the coding priorities will tend to lead to an underestimation of the overall use of natural groups because only those studies that rely solely on this type of data were coded as such.

10 Whitley, *Intellectual and Social Organization.*

11 Sokal, "James McKeen Cattell."

12 In the public bureaucracy of education, with its traditional respect for academic qualifications, the educational psychologists were rather better protected from this threat. They were the designers and evaluators of mental tests, not their administrators. That task was relegated to a low-status stratum of "school psychologists" and teachers whose freedom of action was seriously circumscribed by the bureaucratic context in which they had to operate.

13 In any case there was a fundamental convergence between the kind of social situation in which the now dominant style of psychological experimentation was carried out and the social situation in mental testing. In both cases the situation was marked by an asymmetrical relationship in which the investigator was in a position of control and produced the kind of knowledge that was relevant to those in controlling positions. See the discussion of social practices of investigation in chapter 4. The issue is taken up again in chapter 11.

14 K. Danziger, "Statistical method and research practice in American psychology," in G. Gigerenzer, L. Kruger, and M. Morgan, eds., *The Probabilistic Revolution: Ideas in Modern Science,* vol. 2 (Cambridge, Mass.: MIT Press, 1987), pp. 35–47.

15 J. M. O'Donnell, *The Origins of Behaviorism: American Psychology, 1870–1920* (New York: New York University Press, 1985).

16 Perhaps the most insightful of these critics was Madison Bentley who expressed himself as follows in his mid–1930s autobiographical statement: "These new changes are in part the outgrowths of two decades when pressure to extend psychology toward the practical arts came chiefly from those arts themselves and not from a growing abundance of facts and principles within psychology itself." He goes on to criticize the common practice "of merely translating situations which arise in business, advertising, mental disorder, crime, delinquency, and so on, into terms which possess the flavour instead of the essence of a well-integrated science." See M. Bentley, in Carl Murchison, ed., *A History of Psychology in Autobiography,* vol. 3 (Worcester, Mass.: Clark University Press, 1936), pp. 63–64. A few years before, C. H. Judd had complained in a similar retrospective statement that "psychology is paying the price of its popularity." He counseled that "it will have to be cured of the idea that a hundred loosely related facts can be welded into a body of scientific truth by averaging discordant tendencies and covering up most of the facts by statistical juggling." See C. H. Judd, in Carl Murchison, ed., *A History of Psychology in Autobiography,* vol. 2 (Worcester, Mass.: Clark University Press, 1932), p. 230.

17 O'Donnell, "Crisis of Experimentalism." Boring's role as a critic of the misuse of statistics in psychology is discussed in chapter 9 of the present volume.

18 R. C. Tobey, *The American Ideology of National Science, 1919–1930* (Pittsburgh: University of Pittsburgh Press, 1971).

19 In F. Samelson, "Organizing for the kingdom of behavior: Academic battles and organizational policies in the twenties," *Journal of the History of the Behavioral Sciences* 21 (1985):33–47. The Carnegie Foundation had been financing mental testing on a very large scale and also supported the research of key representatives of the new style of psychological research, like Terman and Thorndike, in a very generous way; see M. Russel, "Testers, trackers and trustees: The ideology of the intelligence testing movement in America 1900–1954" (Ph.D. diss., University of Illinois at Urbana–Champaign, 1972), chap. 3.

20 For an analysis of a particularly important instance of the process by which a formalistic kind of theory emerged "in the shadow of functionalist and pragmatic rhetoric," see J. G. Morawski, "Organizing knowledge and behavior at Yale's Institute of Human Relations," *Isis* 77 (1986):219–242.

21 On the role of industrial psychology in Germany, see S. Jaeger and I. Staeuble, "Die Psychotechnik und ihre gesellschaftlichen Entwicklungsbedingungen," in F. Stoll, ed., *Die Psychologie des 20. Jahrhunderts,* vol. 13 (Zurich: Kindler, 1980), pp. 53–95. Also S. Jaeger, "Zur Herausbildung von Praxisfeldern der Psychologie bis 1933," in M. G. Ash and U. Geuter, eds., *Geschichte der deutschen Psychologie im 20. Jahrhundert* (Opladen: Westdeutscher Verlag, 1985), pp. 83–112.

22 W. Stern, *Die differentielle Psychologie in ihren methodischen Grundlagen* (Leipzig: Barth, 1911); W. Stern, *Die psychologischen Methoden der Intelligenzprüfung und deren Anwendung an Schulkindern* (Leipzig: Barth, 1912); W. Betz, "Ueber Korrelation: Methoden der Korrelationsberechnung und kritischer Bericht über Korrelationsuntersuchungen aus dem Gebiete der Intelligenz, der Anlagen und ihrer Beeinflussung durch äussere Umstände," *Beihefte zur Zeitschrift für angewandte Psychologie und psychologische Sammelforschung* 3 (1911):1–80; G. Deuchler, "Ueber die Methoden der Korrelationsrechnung in der Pädagogik und Psychologie," *Zeitschrift für pädagogische Psychologie und experimentelle Pädagogik* 15 (1914):114–131, 145–159, 229–242. Spearman studied at Leipzig and collaborated with Felix Krueger, who was to become Wundt's successor; see F. Krueger and C. Spearman, "Korrelation zu verschiedenen geistigen Leistungsfähigkeiten," *Zeitschrift für Psychologie* 44 (1906):50–114.

23 P. Lundgreen, *Sozialgeschichte der deutschen Schule im Ueberblick,* Teil 1 (Göttingen: Vandenhoeck & Ruprecht, 1980).

24 P. Lundgreen, ibid., Teil 2 (Göttingen: Vandenhoeck & Ruprecht, 1981).

25 For the specific considerations that underlie this highly synoptic statement, see F. K. Ringer, *Education and Society in Modern Europe* (Bloomington: Indiana University Press, 1979), chap. 5.

26 Mental tests, the main product on this market, were sometimes regarded with suspicion by classroom teachers because, although they sometimes allowed teachers to rationalize their failures, they also diminished the authority of their judgments of ability and because administrators used test results to measure the performance of teachers. See P. D. Chapman, "Schools as Sorters: Lewis M. Terman and the Intelligence Testing Movement, 1890–1930" (Ph.D. diss., Stanford University, 1979), pp. 185ff. This source also provides a cross-country comparison of test development (p. 167) which shows the early and overwhelming predominance of the United States in this field.

27 On the foundation of the Leipzig research institute and its significance for educational research in Germany, see the following: K. Ingenkamp, "Das Institut des Leipziger Lehrervereins 1906–1933 und seine Bedeutung für die Empirische Pädagogik," *Empirische Pädagogik* 1 (1987):60–70; C. C. Kohl, "The Institute for Experimental Pedagogy and Psychology of the Leipzig Lehrerverein," *Journal of Educational Psychology* 4 (1913):367–369; also P. Schlager, "Das Institut für experimentelle Pädagogik und Psychologie," *Pädagogisch-Psychologische Arbeiten* 1

(1910):V–XïI; and a report in *Zeitschrift für experimentelle Pädagogik* 7 (1908):218–223. Vol. 10 (1910):133–139 of the same journal contains a report on the Berlin Institute. Subsequent volumes of *Pädagogisch-Psychologische Arbeiten* and of the *Zeitschrift für pädagogische Psychologie und experimentelle Pädagogik* contain regular reports on the activities of these and other similar institutes, including their budgets and lists of their supporters.

28 The Leipzig teachers' association published its own research journal, *Pädagogisch-Psychologische Arbeiten,* from 1910. Of the twenty empirical studies published there between 1910 and 1916 only one relied predominantly on group data, while eleven relied mainly on individual data, the rest being of the mixed type. In the 1920s group data studies reach half of the total.

29 Prominent examples of this pattern are provided by Karl Marbe and William Stern. See their autobiographical statements in Carl Murchison, ed., *A History of Psychology in Autobiography,* vol. 1 (Worcester, Mass.: Clark University Press, 1930), pp. 335–388, and vol. 3 (Worcester, Mass.: Clark University Press, 1936), pp. 181–213. The ambiguity of Stern's position when faced with practical questions emerges rather clearly in W. Stern, "Der personale Faktor in Psychotechnik und praktischer Psychologie," *Zeitschrift für angewandte Psychologie* 44 (1933):52–63; for an overview in English there is his *General Psychology from the Personalistic Standpoint* (New York: Macmillan, 1938); I. Staeuble, "William Stern's research program of differential psychology: Why did psychotechnics outstrip psychognostics?" a paper read at the 15th annual meeting of the Cheiron Society, Toronto, 1983.

30 See M. G. Ash, "Academic politics in the history of science: Experimental psychology in Germany, 1879–1941," *Central European History* 13 (1981):255–286. On the contrast between American and German psychology in this respect, see K. Danziger, "The social origins of modern psychology," in A. R. Buss, ed., *Psychology in Social Context* (New York: Irvington, 1979), pp. 27–45.

31 The terms in which an influential figure like E. Spranger opposed standardized tests were characteristic: "for the individual is viewed here in the end as something measurable and graspable in numbers, not as a structural principle of the soul." Quoted in M. G. Ash, "Psychology in twentieth-century Germany: Science and profession," in G. Cocks and K. Jarausch, eds., *German Professions 1800–1950* (New York: Oxford University Press, 1990).

32 The situation in psychology had its parallels in other disciplines. See J. Harwood, "National styles in science: Genetics in Germany and the United States between the World Wars," *Isis* 78 (1987):390–414.

33 D. Ruschemeyer, "Professionalisierung: Theoretische Probleme für die vergleichende Geschichtsforschung," *Geschichte und Gesellschaft* 6 (1980):311–325.

Chapter 9. From quantification to methodolatry

1 Cf. chapter 5, n. 14.

2 G. T. Fechner, *Elemente der Psychophysik* (Leipzig: Breitkopf & Härtel, 1860) and *In Sachen der Psychophysik* (Leipzig: Breitkopf & Härtel, 1877).

3 There is a comprehensive discussion of the issue, as it presented itself to the classical experimentalists in psychology, in E. B. Titchener, *Experimental Psychology: A Manual of Laboratory Practice* (New York: Macmillan, 1905), vol. 2, pt. 2. See also S. M. Stigler, *The History of Statistics: The Measurement of Uncertainty before 1900* (Cambridge, Mass.: Harvard University Press), pp. 253–254.

4 On Fechner's commitment to physical and mental atomism, see M. E. Marshall, "Physics, metaphysics, and Fechner's psychophysics," in W. R. Woodward and M. G. Ash, eds., *The Problematic Science: Psychology in Nineteenth-Century Thought* (New York: Praeger, 1982), pp. 65–87.

5 Much of the early literature on these so-called *Vexirfehler,* or paradoxical judgments, is reviewed in E. G. Boring, *Sensation and Perception in the History of Experimental Psychology* (New York: Appleton-Century-Crofts, 1942), pp. 479–480.

6 For a perceptive account of Binet's critical contributions to this issue, see T. H. Wolf, *Alfred Binet* (Chicago: University of Chicago Press, 1973), pp. 110–115.

7 See the series of studies on the two-point threshold published by Alfred Binet in volume 9 of *L'Année psychologique.* A translated excerpt from these has been published in R. H. Pollack and M. W. Brenner, eds., *The Experimental Psychology of Alfred Binet: Selected Papers* (New York: Springer, 1969), pp. 13–27. Prior to this the problems had been investigated quite systematically by Tawney, although not with Binet's fundamental critical insight; G. A. Tawney, "Ueber die Wahrnehmung zweier Punkte mittelst des Tastsinnes, mit Rücksicht auf die Frage der Uebung und die Entstehung der Vexirfehler," *Philosophische Studien* 13 (1897):163–221, and "Ueber die Trugwahrnehmung zweier Punkte bei der Berührung eines Punktes der Haut," *Philosophische Studien* 11 (1895):394–405.

8 This useful term was introduced by Nicholas Georgescu-Roegen in *The Entropy Law and the Economic Process* (Cambridge, Mass.: Harvard University Press, 1971), chap. 2, to refer to discrete and nonoverlapping concepts. I use the term only to refer to such concepts when they are also serially arranged.

9 This particular series is due to E. J. Gates, "The determination of the limens of single and dual impression by the method of constant stimuli," *American Journal of Psychology* 26 (1915):152–157; references to other, more complicated variants will be found in Boring, *Sensation and Perception,* p. 481.

10 E. B. Titchener, "On ethnological tests of sensation and perception with special reference to the tests of color vision and tactile discrimination described in the reports of the Cambridge Anthropological Expedition to Torres Straits," *Proceedings of the American Philosophical Society* 55 (1916):204–236.

11 Boring, *Sensation and Perception,* p. 482.

12 The best-known examples were those offered by William James: "To introspection, our feeling of pink is surely not a portion of our feeling of scarlet; nor does the light of an electric arc seem to contain that of a tallow-candle in itself" (*Principles of Psychology,* vol. 1 [New York: Henry Holt, 1890], p. 546). In his criticism of psychophysics published in 1887, the widely read philosopher, Henri Bergson, had generalized thus: "The fact is that there is no point of contact between the unextended and the extended, between quality and quantity. We can interpret the one by the other, set up the one as equivalent of the other; but sooner or later, at the beginning or at the end, we shall have to recognize the conventional character of the assimilation." See *Essai sur les données immédiates de la conscience,* translated as *Time and Free Will* in 1910 (New York: Harper Torchbook edition, 1960), p. 70. Also, W. W. Meissner, "The problem of psychophysics: Bergson's critique," *Journal of General Psychology* 66 (1962):301–309. Natural scientists also had their doubts; see J. von Kries, "Ueber die Messung intensiver Grössen und über das sogenannte psychophysische Gesetz," *Vierteljahresschrift für wissenschaftliche Philosophie und Soziologie* 6 (1882):257–294.

13 The classical review of this literature is E. G. Boring, "The stimulus error,"

American Journal of Psychology 32 (1921):449–471, reprinted in E. G. Boring, *History, Psychology, and Science: Selected Papers* (New York: Wiley, 1963). For a modern reconsideration, see G. A. Hornstein, "Quantifying psychological phenomena: Debates, dilemmas and implications," in J. G. Morawski, ed., *The Rise of Experimentation in American Psychology* (New Haven, Conn.: Yale University Press, 1988), pp. 1–34. When psychologists began to measure *attitudes* as well as sensations, the classical objections resurfaced, though without any practical effect; see H. M. Johnson, "Pseudo-mathematics in the mental and social sciences," *American Journal of Psychology* 48 (1936):342–351.

14 On the general problem of "measurement by fiat" in the social sciences, see A. V. Cicourel, *Method and Measurement in Sociology* (Glencoe, Ill.: The Free Press, 1964), chap. 1.

15 H. Ebbinghaus, *Memory: A Contribution to Experimental Psychology*, trans. H. A. Ruger and C. E. Bussenius (New York: Teachers College, Columbia University, 1913), from the German original, 1885.

16 That there was common ground between Wundt and Ebbinghaus emerges quite clearly when one compares their methods with other early attempts at an empirical study of memory: e.g., F. Galton, "Psychometric experiments," *Brain* 2 (1879):149–162; J. Delboeuf, "La mémoire chez les hypnotisées," *Revue philosophique* 21 (1886):441–472. The ideal of experimental control, both on the stimulus and on the response side, was much more intensively applied by the German investigators. This was not easily combined with an investigation of memory in its nonreproductive aspects.

17 In Wundt's laboratory this function was represented as a logarithmic one with two "constants" whose value varied from individual to individual. K. H. Wolfe, "Untersuchungen über das Tongedächtnis," *Philosophische Studien* 3 (1886): 534–571.

18 W. Wundt, *Grundzüge der physiologischen Psychologie* (Leipzig: Engelmann, 1887), vol. 2, pp. 359–364.

19 K. Danziger, "Hermann Ebbinghaus and the psychological experiment," in W. Traxel, ed., *Ebbinghaus Studien 2* (Passau: Passavia, 1987), pp. 217–224.

20 S. Kvale, "Memory and dialectics: Some reflections on Ebbinghaus and Mao Tse-tung," *Human Development* 18 (1975):205–222.

21 H. Ebbinghaus, *Urmanuskript "Ueber das Gedächtnis,"* with an introduction by W. Traxel (Passau: Passavia, 1983), p. 17. This is the original, unpublished version of Ebbinghaus's monograph on his investigations of memory. On the importance of the serial form rather than the individual nonsense syllable for Ebbinghaus's work, see also E. Scheerer, "Ebbinghaus, Herbart and Hegel; or: Not the syllable was meaningless, but the syllables series," in W. Traxel, ed., *Ebbinghaus Studien 2* (Passau: Passavia, 1987), pp. 9–22.

22 This is much more explicit in the draft version of his monograph on memory (see n. 21) than in the published version.

23 The metaphor of psychological activity as a kind of work was also very important for other nineteenth-century pioneers of modern psychology, especially Fechner, whose notion of psychophysics is inconceivable without it. The historical significance of the energy model is discussed extensively in S. Jaeger and I. Staeuble, *Die gesellschaftliche Genese der Psychologie* (Frankfurt: Campus, 1978), pp. 293–315.

24 E. Meumann, *Abriss der experimentellen Pädagogik* (Leipzig: Teubner, 1912).

25 Ibid., pp. 240–242.

26 Meumann contextualizes his research interests as follows: "In the traditional pedagogy we read a great deal about methods of teaching; but in most cases the pedagogical text-books can tell us nothing about methods of learning. And yet we find ourselves confronted by the very serious question as to whether the efficiency of school-room management may not be increased by systematically improving the pupil's procedure in the act of learning in such a fashion that his learning may be perfected in its technical aspects and accomplished more economically. This question becomes the more pressing in modern times because our courses of study, in their attempt to comply with the increasing requirements of practical life, are becoming more and more exacting in the demands which they make upon the memory tasks of school children." E. Meumann, *The Psychology of Learning: An Experimental Investigation of the Economy and Technique of Memory* (New York: Appleton, 1913), p. xiv. (This is an English translation of the third edition of a German work that had first appeared in 1907.)

27 Meumann, *Psychology of Learning*, pp. xvi–xvii. On the emergence of German "experimental pedagogics" and its relationship to Ebbinghaus's research, see S. Sporer, "Gedächtnis in vitro und in vivo: von Herrmann Ebbinghaus' sinnlosen Silben zur experimentellen Pädagogik und zur Aussagepsychologie," in W. Traxel, ed., *Ebbinghaus Studien 2* (Passau: Passavia, 1987), pp. 107–119.

28 Meumann, *Psychology of Learning*, pp. 232, 372–373. The problem was complicated in practice by the need to take into account the appearance of what was called mental or psychological fatigue. Too much fatigue detracted from the economy of learning. Despite numerous attempts in the early years of the century, the phenomena grouped under the heading of "fatigue" resisted the imposition of a satisfactory numerizing structure. Such attempts were renewed later by Clark Hull, whose neobehaviorist translation of fatigue into "inhibition" did not however lead to any greater success with the measurement problem.

29 There is considerable overlap in the content of Meumann's *Psychology of Learning* and the second volume of Thorndike's *Educational Psychology,* entitled *The Psychology of Learning* (New York: Teachers College, Columbia University, 1913). It is the first volume of Thorndike's text, with its psychobiological orientation, that has no real counterpart in Meumann's work. That orientation established a necessary link for Thorndike, and other American psychologists of his generation, between their day-to-day work and the most grandiose ambitions: "It will, of course, be understood that directly or indirectly, soon or late, every advance in the sciences of human nature will contribute to our success in controlling human nature and changing it to the advantage of the common weal," writes Thorndike in an article on "The contributions of psychology to education," in *The Journal of Educational Psychology* 1 (1910):5–12 (p. 8).

30 In the preface to the second edition of his *Vorlesungen zur Einführung in die experimentelle Pädagogik und ihre psychologischen Grundlagen* (Introductory Lectures on Experimental Pedagogics and Its Psychological Foundations) (Leipzig: Engelmann, 1911), Meumann lists ten teachers' associations (*Lehrervereine*) for which he had prepared this material. On the teachers' associations, see chapter 8 of the present volume.

31 Merle Curti in *The Social Ideas of American Educators* (Paterson, N.J.: Littlefield, Adams, 1961), p. 460, considers that Thorndike's "influence in establishing and popularizing the fact-finding, statistical, and experimental technique in education has been immeasurable."

32 The enormous extension in the application of the learning concept did not however affect the interpretation of learning as a kind of work. The claim that all adaptive behavior was governed by universal "principles of learning" simply had the effect of giving the work model the status of a fundamental biological truth. A secularized Calvinism may have played a role here. "My God does not hear prayers. Work for him," proclaims the character who speaks for Thorndike in a dramatic script he never published. See C. J. Karier, *Scientists of the Mind: Intellectual Founders of Modern Psychology* (Urbana and Chicago: University of Illinois Press, 1986), chap. 3.

33 The quoted statement occurs at the beginning of what Thorndike refers to as a "general *Credo*" in a chapter entitled "The nature, purposes, and general methods of measurements of educational products" that formed part of the *17th Yearbook of the National Society for the Study of Education,* pt. 2 (Bloomington, Ill.: Public School Publishing Co., 1918). Some statements commonly attributed to Thorndike on this issue were in fact simplified restatements by disciples. His own attitude is perfectly conveyed by the following statement: "History records no career, war or revolution that can compare in significance with the fact that the correlation between intellect and morality is approximately .3, a fact to which perhaps a fourth of the world's progress is due." See "Educational diagnosis," *Science* 37 (1913):142.

34 See G. Joncich, *The Sane Positivist: A Biography of Edward L. Thorndike* (Middletown, Conn.: Wesleyan University Press, 1968).

35 According to Thorndike, "the facts and laws of psychology . . . should provide the general basis for the interpretation and explanation of the great events studied by history, the complex activities of civilized society, the motives that control the action of labor and capital, and the causes to which linguistic inventions and modifications are due." See *The Elements of Psychology,* 2d ed. (New York: Seiler, 1907), p. 324.

36 Thorndike consistently advocated the abdication of individual judgment in the face of the authority of experts: "Outside that field [our own expertise] the intelligent procedure for most of us is to refuse to think, spending our energy rather in finding the expert in the case and learning from him." From "The psychology of the half-educated man," *Harper* 140 (April 1920):670.

37 Thorndike's conservative identification of the prevailing social order with the order of nature was extreme, even for his generation; see Joncich, *The Sane Positivist,* and Karier, *Scientists of the Mind.* During the Great Depression he wrote: "When it is not feasible to learn what the consequences of weighting one person's satisfactions more than another's will be, our trustee for humanity will do well to weight the wants of good men more than the same wants of bad men, since there is a probability that the gratification of wants will cause both to maintain or increase their customary activities" (*Science* 83 [1936]:1–8).

38 W. G. Sleight, "Memory and formal training," *British Journal of Psychology* 4 (1911):386–457. Twelve years earlier Binet's associate, Victor Henri, had made a general suggestion about using an estimate of variability to establish the significance of the difference between group averages: V. Henri, "Quelques applications du calcul des probabilités a la psychologie," *L'Année Psychologique* 5 (1899):153–160. However, at that point there was no group of psychological investigators for whose work the suggestion had practical relevance, and so it was ignored and remained buried in the literature. It is not even mentioned in a 1929 bibliography compiled for psychologists and educationists that contains many older and less directly relevant items; see H. Walker, *Studies in the History of Statistical Method* (Baltimore: Williams & Wilkins, 1929).

39 E.g., W. H. Winch, "Experimental researches on learning to spell," *Journal of*

Educational Psychology 4 (1913):525–537 (p. 535). Other investigators also took notice of the statistical weakness of Winch's earlier studies and introduced the appropriate techniques into their own work – e.g., E. J. G. Bradford, "A note on the relation and aesthetic value of the perceptive types in color appreciation," *American Journal of Psychology* 24 (1913):545–554 (p. 548).

40 For an example of graphical methods of group comparison, see W. G. Chambers, "Individual differences in grammar grade children," *Journal of Educational Psychology* 1 (1910):61–75.

41 J. R. McGaughy, *The Fiscal Administration of City School Systems* (New York: Macmillan, 1924).

42 A comparison of two widely used texts of the time with their earlier counterparts indicates that the treatment of this topic had become more prominent and extensive, and that consideration of less mechanical methods for assessing the degree of overlap between two distributions had been dropped. See H. Sorenson, *Statistics for Students of Psychology and Education* (New York: McGraw-Hill, 1936); H. E. Garrett, *Statistics in Psychology and Education,* 2d ed. (New York and London: Longmans, Green, 1937) (1st, 1926); T. L. Kelley, *Statistical Method* (New York: Macmillan, 1924); E. L. Thorndike, *An Introduction to the Theory of Mental and Social Measurements,* 2d ed. (New York: Science Press, 1919).

43 This development probably prepared the ground for the relatively rapid wholesale adoption of Fisherian techniques of significance testing and experimental design in the subsequent period. See A. J. Rucci and R. D. Tweney, "Analysis of variance and the 'second discipline' of scientific psychology: A historical account," *Psychological Bulletin* 87 (1980):166–184. At the same time, the methodological preconceptions formed during the earlier period led to certain difficulties in appreciating the special features of the Fisherian approach. See A. D. Lovie, "The analysis of variance in experimental psychology: 1934–1945," *British Journal of Mathematical and Statistical Psychology* 32 (1979):151–178. The outcome was a curious hybrid of Fisherian and non-Fisherian approaches. See G. Gigerenzer and D. J. Murray, *Cognition as Intuitive Statistics* (Hillsdale, N.J.: Lawrence Erlbaum, 1987), chap. 1.

44 J. O'Donnell, "The crisis of experimentalism in the 1920s: E. G. Boring and his use of history," *American Psychologist* 34 (1979):289–295; and D. A. Stout, "Personal politics or policing a profession? A methodological debate between E. G. Boring and Carl Murchison" (unpublished manuscript). Boring was not the only experimentalist who inveighed against the misuse of statistics encouraged by the predominance of applied interests; see, e.g., K. Dunlap, "The experimental methods of psychology," in M. Bentley et al., *Psychologies of 1925* (Worcester, Mass.: Clark University, 1926).

45 E. G. Boring, "Mathematical vs. scientific significance," *Psychological Bulletin* 16 (1919):335–338; "The logic of the normal law of error in mental measurement," *American Journal of Psychology* 31 (1920):1–33; "Scientific induction and statistics," *American Journal of Psychology* 37 (1926):303–307.

46 Boring, "Significance," p. 338.

47 Boring, "Logic," p. 33.

48 T. L. Kelley, "The principles and techniques of mental measurement," *American Journal of Psychology* 34 (1923):408–432 (p. 419).

49 The most explicit of these conventions concerned "acceptable" levels of statistical significance, which appear to have been extensions of older conventions. See M. Cowles and C. Davis, "On the origin of the .05 level of statistical significance,"

American Psychologist 37 (1982):553–558. Such conventions defining "reasonable agreement" are not of course peculiar to psychology; see T. S. Kuhn, "The function of measurement in modern physical science," *Isis* 52 (1961):162–176.

50 In later years some critical voices were raised against this and related practices. See D. E. Morrison and R. E. Henkel, eds., *The Significance Test Controversy* (Chicago: Chicago University Press, 1970).

51 On the significance of these practices see G. Gigerenzer, "Probabilistic thinking and the fight against subjectivity," in L. Krüger, G. Gigerenzer, and M. S. Morgan, eds., *The Probabilistic Revolution,* vol. 2.: *Ideas in the Sciences* (Cambridge, Mass.: MIT Press, 1987); also Gigerenzer and Murray, *Intuitive Statistics,* chap. 1; and M. C. Acree, "Theories of Statistical Inference in Psychological Research: A Historico-Critical Study" (Ph.D. diss., Clark University, 1978).

52 B. D. Mackenzie, *Behaviourism and the Limits of Scientific Method* (London: Routledge and Kegan Paul, 1977).

53 D. Bakan, *On Method* (San Francisco: Jossey-Bass, 1967) chaps. 1–3. This practice has also been troubling to that minority of psychologists who have perceived the inconsistency between it and traditional scientific determinism; see M. Sidman, *Tactics of Scientific Research* (New York: Basic Books, 1960).

54 W. S. Hunter, "Correlation studies with the maze in rats and humans," *Comparative Psychology Monographs* 1 (1922), no. 1, pt. 2; W. S. Hunter and R. Vance, "Further studies on the reliability of the maze with rats and humans," *Journal of Comparative Psychology* 4 (1924):431–442; H. Carr, "The reliability of the maze experiment," *Journal of Comparative Psychology* 6 (1926):85–93; W. S. Hunter, "Reply to Professor Carr on 'The reliability of the maze experiment,'" *Journal of Comparative Psychology* 6 (1926):393–398; R. Tryon, "Effect of the unreliability of measurement on the difference between groups," *Journal of Comparative Psychology* 6 (1926):449–453; E. C. Tolman and D. B. Nyswander, "The reliability and validity of maze measures for rats," *Journal of Comparative Psychology* 7 (1927):425–460; R. C. Tryon, "Demonstration of the effect of unreliability of measurement on a difference between groups," *Journal of Comparative Psychology* 8 (1928):1–22. The general implications of the discussion were not lost on psychologists working outside this field; see H. R. Crosland, "Certain points concerning the reliabilities of experiments in psychology," *American Journal of Psychology* 40 (1928):331–337.

55 K. Dunlap, "The average animal," *Journal of Comparative Psychology* 19 (1935):1–3.

56 D. L. Krantz, "Schools and systems: The mutual isolation of operant and non-operant psychology as a case study," *Journal of the History of the Behavioral Sciences* 8 (1972):86–102 (p. 92).

57 On the problems of imposing statistical models involving the continuity assumption on distributions that mask an underlying psychological discontinuity, see R. Harré, *Social Being* (Oxford: Blackwell, 1979), p. 108; also M. Oakes, *Statistical Inference: A Commentary for the Social and Behavioural Sciences* (Chichester and New York: Wiley, 1986), pp. 163–166. As pointed out in chapter 5, the issue goes back to the nineteenth-century roots of social statistics.

58 For further discussion of this issue, see K. Danziger, "The methodological imperative in psychology," *Philosophy of the Social Sciences* 15 (1985):1–13.

59 P. E. Meehl, "Theory testing in psychology and physics: A methodological paradox," *Philosophy of Science* 34 (1967):103–115; also Oakes, *Statistical Inference,* chap. 2.

60 To avoid possible misunderstanding I should add that this criticism is directed at the historically pervasive misuse of statistical inference, not at its use in connection with problems that are inherently statistical in nature.

61 K. Danziger, "On theory and method in psychology," in W. Baker, L. Mos, and H. Rappard, eds., *Recent Trends in Theoretical Psychology* (New York: Springer-Verlag, 1988), pp. 87–94.

62 See Gigerenzer and Murray, *Intuitive Statistics,* chaps. 2 and 5.

63 W. Wundt, *Outlines of Psychology* (Leipzig: Engelmann, 1897), sec. 24. A fuller statement appears in his *Logik,* 3d ed., vol. 3 (Stuttgart: Enke, 1908), pp. 274–282.

Chapter 10. Investigating persons

1 F. Samelson, "Putting psychology on the map: Ideology and intelligence testing," in A. R. Buss, ed. *Psychology in Social Context* (New York: Irvington, 1980), pp. 103–168.

2 R. T. von Mayrhauser, "The manager, the medic, and the mediator: The clash of professional psychological styles and the wartime origins of group mental testing," in M. M. Sokal, ed., *Psychological Testing and American Society* (New Brunswick, N.J.: Rutgers University Press, 1987), pp. 128–157.

3 J. R. McCrory, "A study of the relation between ability and achievement," *Educational Administration and Supervision* 12 (1926):481–490 (see p. 490).

4 See, e.g., A. T. Poffenberger and F. L. Carpenter, "Character traits in school success," *Journal of Experimental Psychology* 7 (1924):67–74; W. H. Hughes, "Organized personnel research and its bearing on high-school problems," *Journal of Educational Research* 10 (1924):386–398; G. E. Manson, "Personality differences in intelligence test performance," *Journal of Applied Psychology* 9 (1925):230–255; A. E. Traxler, "The Will–Temperament of upper-grade and high-school pupils," *School Review* 33 (1925):264–273. In 1921 (vol. 12), the *Journal of Educational Psychology* published a symposium on "Intelligence and its measurement." Many of the contributors called for an extension of mental testing to cover character and personality; see, e.g., Freeman, p. 139, Pintner, p. 142, Pressey, p. 147, Haggerty, p. 216, Buckingham, p. 275.

5 The history of intelligence testing had always been marked by a fundamental divergence between those who were guided by concrete practical goals, like Binet, and those with a commitment to the goals of eugenics, like most of the well-known American and British pioneers in the field. The latter tended to give some global significance to that which intelligence tests measured. In one sense the history of the field consists of a series of concessions forced upon the second group by the practical demonstrations of the first group. See, e.g., von Mayrhauser, "Manager, medic, and mediator"; T. M. Wolf, *Alfred Binet* (Chicago: University of Chicago Press, 1973); H. Cravens, *The Triumph of Evolution* (Philadelphia: University of Pennsylvania Press, 1978), chap. 7.

6 For a detailed consideration of these changes, see J. Parker, "From the Intellectual to the Non-intellectual Traits: A Historical Framework for the Development of American Personality Research" (M.A. thesis, York University, Toronto, 1986). I am much indebted to this source for the historical background of the issues discussed in this section.

7 Methodological implications of this practice are critically considered in J. T. Lamiell, *The Psychology of Personality* (New York: Columbia University Press,

1987). See also the classical analysis of the topic in P. Lafitte, *The Person in Psychology: Reality or Abstraction* (London: Routledge and Kegan Paul, 1957).

8 For a comprehensive critical discussion of this issue, see G. Gigerenzer, *Messung und Modellbildung in der Psychologie* (Munich: Reinhardt, 1981), pp. 88–112.

9 As an illustration, the personality category of extraversion–introversion certainly did not owe its popularity to the fact that many American psychologists held Jung's theoretical system in high esteem or were even especially interested in it. Rather, these terms were prised loose from their original theoretical integument and generally assimilated to prevailing popular conceptions of the healthy "well-adjusted" personality and its opposite.

10 J. Parker, "The evolution of trait concepts in American personality research: Trait terms used in the *Journal of Abnormal and Social Psychology,* 1920 to 1955," a paper read at the 6th annual European Cheiron conference, Brighton, 1987.

11 S. W. Fernberger, "The scientific interests and scientific publications of the members of the American Psychological Association," *Psychological Bulletin* 35 (1938):261–281. The year 1937 was marked by the appearance of G. W. Allport's *Personality: A Psychological Interpretation* (New York: Holt, Rinehart & Winston, 1937), an influential text that established the scope and definition of the new field for many years to come.

12 G. W. Allport, "Personality and character," *Psychological Bulletin* 18 (1921):441–455; P. M. Symonds, "The present state of character measurement," *Journal of Educational Research* 15 (1924):484–498.

13 For a comprehensive account, see C. M. Shea, "The Ideology of Mental Health and the Emergence of the Therapeutic Liberal State: The American Mental Hygiene Movement, 1900–1930" (Ph.D. diss., University of Illinois at Urbana-Champaign, 1980).

14 During the 1920s mental hygiene appears to have received about as much money from private foundations as the entire field of psychology. See table 11 in R. E. Kohler, "Science, foundations and American universities in the 1920's," *Osiris* n.s. 3 (1987):135–164.

15 In this connection John Burnham speaks of "the concept of improving the world not by social action but by a patient program of individual treatment of large numbers of people." See J. Burnham, "The new psychology: From narcissism to social control," in J. Braeman, R. H. Brenner, D. Brody, eds., *Change and Continuity in Twentieth Century America: The 1920's* (Columbus: Ohio State University Press, 1968), pp. 351–398. Also Shea, "Ideology of Mental Health."

16 The Laura Spelman Rockefeller Memorial Fund, established in 1918 with an endowment of $74 million, played a crucial role here. See the archival material referred to in A. B. Smuts, "The National Research Council Committee on Child Development and the founding of the Society for Research in Child Development, 1925–1933," in A. B. Smuts and J. W. Hagen, "History and Research in Child Development," *Monographs of the Society for Research in Child Development* 50 (1985):4–5 (serial no. 211).

17 This reference of the term "personality" is to be clearly distinguished from several other meanings the term may take on in everyday discourse. Of all the slippery terms that define modern psychological discourse this one is perhaps the most slippery. On some historical variants, see R. Williams, *Keywords: A Vocabulary of Culture and Society* (London: Fontana, 1976); and W. I. Susman, "Personality and the making of twentieth-century culture," in J. Higham and P. K.

Conkin, eds., *New Directions in American Intellectual History* (Baltimore: Johns Hopkins University Press, 1979), pp. 212–226.

18 In an analysis of the profound influence of the mental hygiene movement on American education Sol Cohen expresses this as follows: "The idea of the school's responsibility for 'personality development' is a kind of ideogram which serves as shorthand for a systematically related cluster of attitudes, assumptions and practices with profound implications for curriculum, grading and promotion practices, methods of instruction, notions of achievement, authority, and discipline, and which also includes a tacit assumption by the schools of a broad parent–surrogate function." S. Cohen, "The school and personality development: Intellectual history," in J. H. Best, ed., *Historical Inquiry in Education: A Research Agenda* (Washington, D.C.: American Educational Research Association, 1983), p. 110. Also see S. Cohen, "The mental hygiene movement, the development of personality and the school: The medicalization of American education," *History of Education Quarterly* 23 (1983):123–149.

19 P. Shermer, "The development of research practice in abnormal and social psychology: 1906–1956" (M.A. thesis, York University, Toronto, 1983). See also R. W. Lissitz, "A longitudinal study of the research methodology in the Journal of Abnormal and Social Psychology, the Journal of Nervous and Mental Disease, and the American Journal of Psychiatry," *Journal of the History of the Behavioral Sciences* 5 (1969):248–255. The *Journal of Personality* (at first *Character and Personality*), which began to appear in the 1930s, showed a preference for statistical studies almost immediately. For more recent trends, see K. H. Craik, "Personality research methods: An historical perspective," in *Journal of Personality* 54 (1986):18–51.

20 See L. Cochran, "On the categorization of traits," *Journal for the Theory of Social Behaviour* 14 (1984):183–209.

21 P. E. Vernon, "The American v. the German methods of approach to the study of temperament and personality." *British Journal of Psychology* 24 (1933): 156–177.

22 W. Dilthey, "Ideen über eine beschreibende und zergliedernde Psychologie," *Sitzungsberichte der Akademie der Wissenschaften zu Berlin* 2 (1894):1309–1407. (Reprinted in W. Dilthey, *Gesammelte Schriften,* vol. 5 [Stuttgart: Teubner, 1957], pp. 139–240.)

23 T. Lipps, *Psychologische Untersuchungen I* (Leipzig: Engelmann, 1907).

24 A few years previously Wundt had presented a somewhat similar theory; see W. Wundt, *Völkerpsychologie,* vol. 1 (Leipzig: Engelmann, 1900), and K. Danziger, "Origins and basic principles of Wundt's Völkerpsychologie," *British Journal of Social Psychology* 22 (1983):303–313.

25 Lipp's views were summarized in his famous *empathy* concept (Einfühlung), a term that experienced some strange transmutations of meaning after its adoption by American psychologists. See K. Holzkamp, "Zur Geschichte und Systematik der Ausdruckstheorien," in R. Kirchkoff, ed., *Ausdruckspsychologie* (Göttingen: C. J. Hogrefe, 1965), pp. 39–113; also G. A. Gladstein, "The historical roots of contemporary empathy research," *Journal of the History of the Behavioral Sciences* 20 (1984):38–59.

26 L. Klages, *Die Grundlegung der Wissenschaft vom Ausdruck* (1907; repr. Bonn: Bouvier, 1964); and L. Klages, *Handschrift und Charakter* (1917; repr. Bonn: Bouvier 1968). The latter book enjoyed an extraordinary popularity for half a century, the 1968 edition being its twenty-sixth.

27 Graphology as a systematic method had emerged at the same time as experimental psychology. The term is said to have made its first appearance in J. H. Michon, *Système de Graphologie*, (Paris: Marpon et Flammarion, 1875).

28 See R. Kichhoff, "Zur Geschichte des Ausdrucksbegriffs" (1965), in Kirchhoff, *Ausdruckspsychologie*; also K. Bühler, *Ausdruckstheorie* (Jena: G. Fischer, 1933).

29 This type of study is pursued in contemporary social psychology under the heading of "nonverbal communication."

30 The idea of "character" as a knowledge object to be accessed in a systematic and methodical way originates with J. Bahnsen, *Beiträge zur Characterologie mit besonderer Berücksichtigung pädagogischer Fragen* (Leipzig: Brockhaus, 1867). The practical context for this development was provided by the field of education.

31 See M. G. Ash "Academic politics in the history of science: Experimental psychology in Germany, 1879–1941," *Central European History* 13(1981):255–286.

32 G. Lukacz, *The Destruction of Reason,* trans. P. Palmer (London: Merlin, 1980), from the German original, 1954.

33 See S. Jaeger and I. Staeuble, "Die Psychotechnik und ihre gesellschaftlichen Entwicklungsbedingungen," in F. Stoll, ed., *Die Psychologie des 20. Jahrhunderts,* vol. 13 (Zurich: Kindler, 1980), pp. 53–95.

34 By 1929 there were eight candidates for every vacancy. After that the discrepancy increased. See L. von Renthe-Fink, "Von der Heerespsychotechnik zur Wehrmachtpsychologie" (1985) in *Deutsche Wehrmachtpsychologie 1914–1945* (München: Verlag für Wehrwissenschaften, 1985), p. 87. On the history of military psychology in Germany, see U. Geuter, "Polemos panton pater – Militär und Psychologie im Deutschen Reich 1914–1945," in M. G. Ash and U. Geuter, eds., *Geschichte der deutschen Psychologie im 20 Jahrhundert* (Opladen: Westdeutscher Verlag, 1985), pp. 146–171.

35 In actual fact, the introduction of psychological selection methods was correlated with a distinct *increase* in the proportion of successful candidates originating in the traditional strata from which officers had always been recruited. See D. Bald, "Sozialgeschichte der Rekrutierung des deutschen Offizierkorps von der Reichsgründung bis zur Gegenwart," *Schriftenreihe Innere Führung* (Reihe Ausbildung und Bildung) 29 (München: Sozialwissenschaftliches Institut der Bundeswehr, 1977).

36 For some details see H. L. Ansbacher, "German military psychology," *Psychological Bulletin* 38 (1941):370–392 (p. 383). This paper contains an excellent bibliography, although its own summary gives too much prominence to some aspects given little weight in the originals.

37 Von Renthe-Fink, "Heerespsychotechnik," p. 62.

38 The most influential of the systems of personality psychology developed by German army psychologists was that presented in P. Lersch, *Aufbau des Charakters* (Leipzig: Barth, 1938). His special practical field was the diagnostic use of facial expression. See his *Gesicht und Seele: Grundlinien einer mimischen Diagnostik* (1931; repr. München: Reinhardt, 1966). Other prominent academic psychologists for whose work the military experience was crucial were J. Rudert and K. Mierke.

39 See von Renthe-Fink, "Heerespsychotechnik."

40 This is a major conclusion of the very thorough study of the relationship between psychology and National Socialism presented in U. Geuter, *Die Professionalisierung der deutschen Psychologie im Nationalsozialismus* (Frankfurt: Suhrkamp, 1984).

41 See ibid., pp. 180–186.

42 W. Fritscher, "Die psychologische Auswahl des Offiziernachwuchses in der deutschen Wehrmacht" (1985), in *Deutsche Wehrmachtspsychologie*, pp. 421–475ff.

43 When the selection of an elite became part of the task of Allied psychologists during World War II, they adopted some key features of the German methods. See OSS Assessment Staff, *Assessment of Men* (New York: Rinehart, 1948).

44 Psychological procedures for the selection of officers were abolished by the German army in 1942, largely as a result of enormous losses on the Eastern Front and a drying up of the manpower reservoir. It now seemed more rational to rely on battle testing, rather than on psychological testing, to pick men who would do well under combat conditions.

45 See Geuter, *Die Professionalisierung*, pp. 202–205.

46 There were of course other types of innovative investigative practice in neighboring areas that did not form part of the discipline of psychology. Psychoanalysis and the methods of social research practiced by some members of the Frankfurt school are notable instances. On the latter, see E. Fromm, *The Working Class in Weimar Germany: A Psychological and Sociological Study* (Cambridge, Mass.: Harvard University Press, 1984).

47 Participating as an experimental subject in the investigations of other members of the group was part of the normal experience of those who belonged to this circle. See, e.g., Anita Karsten's autobiographical statement in L. J. Pongratz, W. Traxel, and E. G. Wehner, eds., *Psychologie in Selbstdarstellungen*, vol. 2 (Bern: Huber, 1977), pp. 77–108.

48 T. Dembo, "Der Aerger als dynamisches Problem," *Psychologische Forschung* 15 (1931):1–144 (pp. 73–84).

49 The distinction that is sometimes made between an earlier phase of Lewin's research, directed at personality, and a later phase, directed at social psychology, can be misleading. Most of the "later" social psychological concepts, such as "power fields," occur in the earlier work, as a glance at the research papers published by Lewin's students will show.

50 K. Lewin, "Die Erziehung der Versuchsperson zur richtigen Selbstbeobachtung und die Kontrolle psychologischer Beschreibungsangaben," in C. F. Graumann and A. Métraux, eds., *Kurt Lewin Werkausgabe* (Bern: Huber, and Stuttgart: Klett, 1981), pp. 153–212. This is a paper that Lewin wrote around 1918 but never published in his lifetime.

51 E.g., B. Zeigarnik, "Das Behalten erledigter und unerledigter Handlungen," *Psychologische Forschung* 9 (1927):1–85 (pp. 18–19).

52 K. Lewin, "Vorbemerkungen über die psychischen Kräfte und Energien und über die Struktur der Seele," *Psychologische Forschung* 7 (1926):294–329 (p. 304). It should be noted that the word *Versuch,* which occurs several times in this passage, could mean either an experiment or a trial. I have been guided by the general context of the article in translating this passage.

53 In his later writings Lewin elaborates this point into a full-blown "field theory"; see, e.g., his "Behavior and development as a function of the total situation" in K. Lewin, *Field Theory in Social Science* (London: Tavistock, 1952), pp. 238–303.

54 K. Lewin, "Die psychischen Kräfte," pp. 306–307.

55 N. Ach, *Ueber den Willensakt und das Temperament* (Leipzig: Quelle & Meyer, 1910). For an extensive review of the introspective experimental study of volition, see J. Lindworsky, *Der Wille: Seine Erscheinung und seine Beherrschung nach den*

Ergebnissen der experimentellen Forschung (Leipzig: Barth, 1923). A modern introduction is provided by H. Gundlach, "Anfänge der experimentellen Willenspsychologie," in H. Heckhausen, ed., *Jenseits des Rubikon: Der Wille in den Humanwissenschaften* (Berlin, Heidelberg: Springer-Verlag, 1987), pp. 67–85.

56 K. Lewin, *Die Entwicklung der experimentellen Willenspsychologie und die Psychotherapie,* (Leipzig: Hirzel, 1929), p. 12.

57 In view of Lewin's explicit rejection of phenomenology, a description of his Berlin research program as "experimental phenomenology" seems rather inappropriate. Apart from this unfortunate choice of terminology, the editor's comments in J. de Rivera, *Field Theory as Human-Science: Contributions of Lewin's Berlin Group* (New York: Gardner, 1976), provide a very useful English-language introduction to the topic.

58 Some of his work in this area was reflected in his dissertation, republished as vol. 2 of his collected works. See C. F. Graumann and A. Métraux, eds., *Kurt Lewin Werkausgabe,* vol. 2 (Bern: Huber, and Stuttgart: Klett, 1983).

59 On Lewin's distinction between genotype and phenotype, see "The conflict between Aristotelian and Galilean modes of thought in contemporary psychology," in K. Lewin, *Dynamic Theory of Personality* (New York: McGraw-Hill, 1935), p. 11.

60 Lewin speaks of "the old, mistaken theory of induction" in discussing the nature of psychological experiments. See K. Lewin, "Gesetz und Experiment in der Psychologie," *Symposion* 1 (1927):375–421 (p. 384), republished in Graumann and Métraux, *Kurt Lewin Werkausgabe,* vol. 1, pp. 279–320.

61 K. Lewin, "Gesetz und Experiment," p. 291; Lewin discusses the same point in slightly different terms in "Aristotelian and Galilean modes of thought."

62 Lewin, "Aristotelian and Galilean modes of thought."

63 K. Lewin, "Gesetz und Experiment," p. 406.

64 By far the best English-language source of information on these studies is to be found in de Rivera, *Field Theory as Human-Science*

65 See D. Cartwright, "Theory and practice," *Journal of Social Issues* 34 (1978):168–180. The fate of Lewin's psychology after his death represents a very mixed picture and goes beyond the bounds of this volume. For some positive developments, see E. Stivers and S. Wheelan, eds., *The Lewin Legacy: Field Theory in Current Practice* (Berlin: Springer-Verlag, 1986).

Chapter 11. The social construction of psychological knowledge

1 For various elaborations of this general scheme, see R. Whitley, *The Intellectual and Social Organization of the Sciences* (Oxford: Clarendon Press, 1984); also R. Rilling, *Theorie und Soziologie der Wissenschaft* (Frankfurt: Fischer, 1975).

2 This process is extensively described in B. Latour, *Science in Action* (Cambridge, Mass.: Harvard University Press, 1987).

3 An earlier version of this analysis, applied to the origins of modern psychology, is to be found in K. Danziger, "The social origins of modern psychology," in A. R. Buss, ed., *Psychology in Social Context* (New York: Irvington, 1979), pp. 27–45.

4 W. van den Daele, "Die soziale Konstruktion der Wissenschaft-Institutionalisierung und Definition der positiven Wissenschaft in der zweiten Hälfte des 17. Jahrhunderts," in G. Böhme, W. van den Daele, W. Krohn, *Experimentelle Phi-*

losophie: Ursprünge autonomer Wissenschaftsentwicklung (Frankfurt: Suhrkamp, 1977), pp. 129–184.

5 For specific examples from fields other than psychology see Latour, *Science in Action*.

6 The category of "folk science," introduced by J. R. Ravetz, fits this case exactly: "In each case, the attempt is to reproduce what is believed to be the crucial feature of an established science, where this is learned more from the philosophers of science than from the successful practitioners themselves." See J. R. Ravetz, *Scientific Knowledge and Its Social Problems* (New York: Oxford University Press, 1971), p. 368. Folk science is marked by the "pretence of maturity," and extreme responsiveness to external pressures. It should be added that "folk science" is an analytic category, and there is no implication that all branches of a heterogeneous field like psychology fit this category.

7 D. S. Napoli, *The Architects of Adjustment: The History of the Psychological Profession in the United States* (Port Washington, N.Y.: Kennikat Press, 1981).

8 For some key examples see F. Samelson, "Putting psychology on the map: Ideology and intelligence testing," in A. R. Buss, ed., *Psychology in Social Context* (New York: Irvington, 1980), pp. 103–168; F. Samelson, "Organizing for the kingdom of behavior: Academic battles and organizational policies in the twenties," *Journal of the History of the Behavioral Sciences* 21 (1985):33–47; J. Morawski, "Organizing knowledge and behavior at Yale's Institute of Human Relations," *Isis* 77 (1986):219–242.

9 Several pertinent examples are summarized in S. Shapin, "History of science and its sociological reconstructions," *History of Science* 20 (1982):157–211. A particularly relevant example is discussed by D. Mackenzie and B. Barnes, "Scientific judgement: The Biometry–Mendelism controversy," in B. Barnes and S. Shapin, eds., *Natural Order: Historical Studies of Scientific Culture* (Beverly Hills, Calif.: Sage, 1979), pp. 191–208.

10 In other words, there is something artificial about the distinction between the constitution of a cognitive field and the constitution of the group identified with this field. See, e.g., M. Callan, "Struggles and negotiations to define what is problematic and what is not: The sociologic translation," in K. D. Knorr, R. Krohn, and R. Whitley, eds., *The Social Process of Scientific Investigation* (Dordrecht: Reidel, 1980), pp. 197–219.

11 The inseparable link of ethical and methodological issues has emerged in contemporary discussions of research ethics – e.g., J. G. Adair, T. W. Dushenko, and R. C. L. Lindsay, "Ethical regulations and their impact on research practice," *American Psychologist* 40 (1985):59–72; D. Baumrind, "Research using intentional deception: Ethical issues revisited," *American Psychologist* 40 (1985):165–174; H. Schuler, *Ethical Problems in Psychological Research* (New York: Academic Press, 1982); A. Métraux, "Die Verschränkung methodologischer und ethischer Handlungsregeln," in L. Kruse and M. Kumpf, eds., *Psychologische Grundlagenforschung: Ethik und Recht* (Bern: Huber, 1981), pp. 139–162; B. Harris, "Key words: A history of debriefing in social psychology," in J. G. Morawski, ed., *The Rise of Experimentation in American Psychology* (New Haven: Yale University Press, 1988), pp. 188–212.

12 J. G. Morawski, "Impossible experiments and practical constructions: The social bases of psychologists' work," in Morawski, *The Rise of Experimentation in American Psychology*, pp. 72–93. J. G. Adair and B. Spinner, "Subjects' access to cognitive processes: Demand characteristics and verbal report," *Journal for the*

Theory of Social Behaviour 11 (1981):31–52; J. G. Adair, "Phenomenology and experimental research," a paper presented at the annual meeting of the Canadian Psychological Association, Vancouver, 1987.

13 S. Shapin and S. Schaffer, *Leviathan and the Air-Pump: Hobbes, Boyle and the Experimental Life* (Princeton: Princeton University Press, 1985). On the link between the modern concept of evidence and the social character of witnesses, see also B. J. Shapiro, *Probability and Certainty in Seventeenth-Century England* (Princeton: Princeton University Press, 1983), chap. 5.

14 See E. Freidson, *Professional Powers: A Study of the Institutionalization of Formal Knowledge* (Chicago: University of Chicago Press, 1986); and T. L. Haskell, ed., *The Authority of Experts* (Bloomington: Indiana University Press, 1984).

15 Many facets of this development have been critically described in the writings of Sigmund Koch. See "The nature and limits of psychological knowledge: Lessons of a century qua Science," in S. Koch and D. E. Leary, eds., *A Century of Psychology as Science* (New York: McGraw-Hill, 1985), pp. 75–97, and other papers by this author listed in the above, pp. 96–97.

16 N. Rose, *The Psychological Complex: Psychology, Politics and Society in England 1869–1939* (London: Routledge and Kegan Paul, 1985).

17 A salient example is provided by historical changes in the vogue of hereditarianism and environmentalism among American psychologists. See H. Cravens, *The Triumph of Evolution: American Scientists and the Heredity–Environment Controversy 1900–1941* (Philadelphia: University of Pennsylvania Press, 1978). It should be noted that in psychology swings of the heredity–environment pendulum were made possible by the fact that both positions shared a common ideological basis; see J. Harwood, "Heredity, environment, and the legitimation of social policy," in Barnes and Shapin, *Natural Order,* pp. 231–248. A related change of focus occurred in social psychology; see F. Samelson, "From 'race psychology' to 'studies in prejudice': Some observations on the thematic reversal in social psychology," *Journal of the History of the Behavioral Sciences* 14 (1978):265–278.

18 There are many accounts of the cultural and historical significance of this myth. An analysis that is of particular interest in the present context is to be found in the first three chapters of R. M. Unger, *Knowledge and Politics* (New York: Free Press, 1975). For a discussion of the psychological aspects, see C. Venn, "The subject of psychology," in J. Henriques, W. Hollway, C. Urwin, C. Venn, V. Walkerdine, *Changing the Subject* (London: Methuen, 1984), pp. 119–152.

19 For examples, see E. E. Sampson, "Psychology and the American ideal," *Journal of Personality and Social Psychology* 35 (1977):767–782; and further analysis in E. E. Sampson, "Deconstructing psychology's subject," *The Journal of Mind and Behavior* 4 (1983):135–164.

20 R. Harré, *Personal Being* (Cambridge, Mass.: Harvard University Press, 1984); A. Pepitone, "Lessons from the history of social psychology," *American Psychologist* 36 (1981):972–985.

21 K. J. Gergen, *Toward Transformation in Social Knowledge* (New York: Springer, 1982); R. Harré and P. F. Secord, *The Explanation of Social Behaviour* (Oxford: Blackwell, 1972); K. Holzkamp, "Selbsterfahrung und wissenschaftliche Objektivität: Unaufhebbarer Widerspruch?" in K.-H. Brown and K. Holzkamp, eds., *Subjektivität als Problem psychologischer Methodik* (Frankfurt and New York: Campus, 1985), pp. 17–37.

22 J. D. Greenwood, "On the relation between laboratory experiments and social behavior: Causal explanation and generalization," *Journal for the Theory of Social*

Behaviour 12 (1982):225–250. Analogous problems arise in the currently popular psychology of cognition; see J. Lave, *Cognition in Practice* (New York: Cambridge University Press, 1988).

23 For example, intelligence tests are best at predicting performance in scholastic settings, group performance tasks do relatively well at predicting in group problem-solving situations; in general, what psychological tests do best is predict performance on other psychological tests. On the problems of psychological prediction, see L. J. Cronbach, "Beyond the two disciplines of scientific psychology," *American Psychologist* 30 (1975):116–127. On the question of the relationship between the investigative situation and other social situations, see K. Holzkamp, "Zum Problem der Relevanz psychologischer Forschung für die Praxis," *Psychologische Rundschau* 21 (1970):1–22.

24 The classical statement of the distinction within the context of American psychology is G. W. Allport, *Personality: A Psychological Interpretation* (New York: Holt, Rinehart & Winston, 1937). It should be noted that, although the terms were borrowed from the neo-Kantian philosopher, Windelband, their reference had undergone a characteristic change, the original historical context having been replaced by an ahistorical one.

25 See B. Latour, "Give me a laboratory and I will raise the world," in K. Knorr-Cetina and M. Mulkay, eds., *Science Observed: Perspectives on the Social Study of Science* (Beverly Hills, Calif.: Sage, 1983), pp. 141–170.

26 B. F. Skinner, *Walden Two* (New York: Macmillan, 1948). The distinctive nature of Skinner's utopia is that it is *imposed*. In the real world, psychology may participate in the reconstruction of social settings through its subtle influence on the social consciousness of participants. See A. MacIntyre, "How psychology makes itself true or false," in Koch and Leary, *A Century of Psychology,* pp. 897–903.

27 This delay is all the more remarkable because the point had been made in a prominent journal in 1933. See S. Rosenzweig, "The experimental situation as a psychological problem," *Psychological Review* 40 (1933):337–354. Subsequently, about three decades elapsed before anyone else was ready to take up the issue in print.

28 For an in vivo study of this ideology, see J. Potter and M. Mulkay, "Making theory useful: Utility accounting in social psychologists' discourse," *Fundamenta Scientiae* 3 (1982):259–278.

29 M. Foucault, *Discipline and Punish: The Birth of the Prison* (New York: Pantheon, 1978), pt. 3 (from the French original, 1975). For psychology's contributions as "discipline," see N. Rose, "Calculable minds and manageable individuals," *History of the Human Sciences* 1 (1988):179–200.

30 For a detailed demonstration of the dependence of one set of investigative practices on historically contingent industrial institutions, see B. Schwartz, R. Schuldenfrei, and H. Lacey, "Operant psychology as factory psychology," *Behaviorism* 6 (1978):229–254; also S. Kvale, "The technological paradigm of psychological research," *Journal of Phenomenological Psychology* 3 (1973):143–159.

31 "External validity" is defined in D. T. Campbell and J. C. Stanley, *Experimental and Quasi-experimental Designs for Research* (Chicago: Rand McNally, 1963).

32 During the latter part of the period that is the focus of this study, these ideological functions became increasingly concentrated in the systems of neobehaviorism. See L. D. Smith, *Behaviorism and Logical Positivism* (Stanford: Stanford University Press, 1986); J. A. Mills, "Hull's theory of learning: II. A criticism of the theory and its relationship to the history of psychological thought," *Canadian*

Psychological Review 19 (1978):116–127; J. A. Mills, "Some observations on Skinner's moral theory," *Journal for the Theory of Social Behaviour* 12 (1982):140–160; Morawski, "Organizing knowledge."

33 J. Shotter, *Images of Man in Psychological Research* (London: Methuen, 1975).

34 R. Bhaskar, *A Realist Theory of Science* (Atlantic Highlands, N.J.: Humanities Press, 1978). The implications of this position for the human sciences are discussed in R. Bhaskar, *The Possibility of Naturalism* (Atlantic Highlands, N.J.: Humanities Press, 1979); P. T. Manicas, "The human sciences: A radical separation of psychology and the social sciences," in P. F. Secord, ed., *Explaining Human Behavior: Consciousness, Human Action and Social Structure* (Beverly Hills, Calif.: Sage, 1982), pp. 155–173; P. T. Manicas and P. F. Secord, "Implications for psychology of the new philosophy of science," *American Psychologist* 38 (1983):399–413.

35 The domain of the real in this sense should not be thought of as a domain of entities but as a domain of causal powers. See I. Hacking, *Representing and Intervening* (Cambridge: Cambridge University Press, 1983), chaps. 2 and 12; also R. Harré and E. H. Madden, *Causal Powers* (Totowa, N.J.: Rowman and Littlefield, 1975).

36 The specifics of Wundt's notion of psychic causality cannot be explored here. See his statements in his *Grundzüge der physiologischen Psychologie*, 5th ed., vol. 3 (Leipzig: Engelmann, 1903), and his *Logik*, 3d ed., vol. 3 (Stuttgart: Enke, 1908), pp. 243–295. For an earlier statement in English, see his *Outlines of Psychology* (1897), sec. 5, republished in R. W. Rieber, ed., *Wilhelm Wundt and the Making of a Scientific Psychology* (New York: Plenum, 1980). The significance of the concept of psychic causality for Wundt's experimental practice is discussed in K. Danziger, "Wundt's psychological experiment in the light of his philosophy of science," *Psychological Research* 42 (1980):109–122. See also T. Mischel, "Wundt and the conceptual foundations of psychology," *Philosophy and Phenomenological Research* 31 (1970):1–26.

37 A particularly far-reaching account of Gestalt principles as a form of psychological causality is presented in W. Köhler, *The Place of Value in a World of Facts* (New York: Liveright, 1938).

38 S. Toulmin and D. E. Leary, "The cult of empiricism in psychology, and beyond," in Koch and Leary, *Psychology as Science*, pp. 594–617; K. Danziger, "The positivist repudiation of Wundt," *Journal of the History of the Behavioral Sciences* 15 (1979):205–230; B. D. Mackenzie, *Behaviourism and the Limits of Scientific Method* (Atlantic Highlands, N.J.: Humanities Press, 1977).

39 Scientific activity commonly involves a two-sided procedure: In one phase the empirical domain is accepted and the theoretical domain is put in doubt, but in the other phase the theoretical domain is accepted and the empirical domain is put in doubt. See R. G. Francis, *The Rhetoric of Science* (St. Paul: University of Minnesota Press, 1961). If all the emphasis is placed on either one of these phases, one gets either a jumble of disjointed facts or a list of speculative systems but not a developing science.

40 As Boyd points out, in science "theoretical considerations are evidential." See R. N. Boyd, "The current status of scientific realism," in J. Leplin, ed., *Scientific Realism* (Berkeley: University of California Press, 1984), p. 61.

41 Because alternative conceptions of psychological reality are involved in different types of investigative practice, and therefore lead to different types of data, the confrontation of investigative practice with a divergent theoretical order can sometimes take the form of a confrontation of empirical data. During the period covered

by the present volume, an important example of this type of confrontation was provided by the comparison of Gestalt psychological experiments on problem solving in primates (W. Köhler, *The Mentality of Apes* (New York: Harcourt, Brace, 1925), with American experiments on cats in puzzle boxes and rats in mazes. Here the empirical confrontation served as the occasion for the critique of certain investigative practices from the perspective of a divergent conception of psychological reality. However, such confrontations in depth were the exception rather than the rule. Typically, the original confrontation was rendered harmless by its transformation into the issue of continuity versus discontinuity in learning that flourished in the 1930s and beyond (K. W. Spence, "The nature of discrimination learning in animals," *Psychological Review* 43 [1936]:427–449). With this transformation the potentially disturbing features of the original critique were successfully avoided and the dominant style of investigative practice could continue to be taken for granted. The issue became trivialized to a form of question that could be investigated within the framework of traditional practices. This was the usual fate of controversies that threatened to upset the methodological applecart.

42 For America's first Ph.D. in psychology, J. Jastrow, psychology was already "the master science of the human group," and "the psychologist has come naturally and without presumption to assume the role of helmsman" (*Piloting Your Life: The Psychologist as Helmsman* [Garden City, N.Y.: Greenberg, 1930], p. xiii); E. L. Thorndike (in 1907!) declared that psychology "supplies or should supply the fundamental principles upon which sociology, economics, history, anthropology, linguistics and the other sciences dealing with human thought and action should be based" (*The Elements of Psychology* [New York: Seiler, 1907] p. 323); J. McKeen Cattell added that "it is also for psychology to determine what does in fact benefit the human race" ("The usefulness of psychology," *Science* 72 [1930]:284–287). Although such arrogance could not command universal assent, it did reflect the confidence of men who were securely grounded in the popular prejudices of their age.

43 P. T. Manicas, *A History and Philosophy of the Social Sciences* (Oxford: Blackwell, 1987), pp. 261–265.

44 The general position that "the interplay between 'criteria', 'rules', 'methodological strictures', and the 'cognitive content' of science is constant and tight" is developed extensively in D. Shapere, *Reason and the Search for Knowledge* (Dordrecht: Reidel, 1984) (quotation on p. 216); see also the discussion of an earlier statement of this view by T. Nickels, "Heuristics and justification in scientific research," in F. Suppe, ed., *The Structure of Scientific Theories* (Urbana: University of Illinois Press, 1974), pp. 571–589.

45 D. Bloor, *Wittgenstein: A Social Theory of Knowledge* (New York: Columbia University Press, 1983), chaps. 3 and 7.

46 G. Lakoff, *Women, Fire and Dangerous Things: What Categories Reveal about the Mind* (Chicago: University of Chicago Press, 1987).

47 R. W. Miller, *Fact and Method* (Princeton: Princeton University Press, 1987); R. Harré, *Varieties of Realism* (Oxford: Blackwell, 1986); M. G. Raskin and H. J. Bernstein, *New Ways of Knowing: The Sciences, Society, and Reconstructive Knowledge* (Totowa, N.J.: Rowman and Littlefield, 1987).

48 K. Danziger, "Psychological objects, practice, and history," *Annals of Theoretical Psychology* 8 (New York: Plenum, in press.).

49 "It is necessary for us to see that of the two main ways in which we may explain the growth of science, one is rather unimportant and the other is important. The first explains science by the accumulation of knowledge: it is like a growing library (or a museum). As more and more books accumulate, so more and more knowledge accumulates. The other explains it by criticism: it grows by a method more revolutionary than accumulation – by a method which destroys, changes, and alters, the whole thing including its most important instrument, the language in which our myths and theories are formulated." See K. R. Popper, *Conjectures and Refutations: The Growth of Scientific Knowledge* (New York: Harper & Row, 1968), p. 129.

Index